Teaching Climate Change for Grades 6–12

Looking to tackle climate change and climate science in your class-room? This timely and insightful book supports secondary science teachers in developing effective curricula around the Next Generation Science Standards (NGSS) by grounding their instruction on the climate crisis. This new edition focuses on meeting teachers where they are in their teaching and learning while tending to various contexts, communities, and cultures to activate educators in understanding and responding to the climate crisis in *this* moment. Chapters offer design and implementation support for 21st-century learning experiences centered around the climate emergency for meaningful engagement. Dr. Lê provides an overview of the teaching shifts needed for the NGSS using climate change as the vehicle of instruction. She also supports climate literacy for students and teachers via urgent topics in climate science and environmental justice. Teachers will also learn how to engage with the complexities of climate change by exploring social, racial, and environmental injustices stemming from the climate crisis that directly impact their students. Examples of successful applications of these learning experiences are new to the second edition, as well as added activities and overall updates to research and data. By anchoring instruction on the climate emergency through an intersectional lens starting with teachers' core beliefs and values, Dr. Lê offers guidance on how educators can activate students as agents of change for their own communities.

Kelley T. Lê is the inaugural Executive Director of the University of California (UC) and California State University (CSU) Environmental and Climate Change Literacy Projects (ECCLPs), USA.

Teaching Climate Change for Grades 6–12

Activating Science Teachers to Take on the Climate Crisis Through NGSS

Second Edition

Kelley T. Lê

Routledge
Taylor & Francis Group

NEW YORK AND LONDON

Designed cover image: © Getty Images / Pramote Polyamate

Second edition published 2025
by Routledge
605 Third Avenue, New York, NY 10158

and by Routledge
4 Park Square, Milton Park, Abingdon, Oxon, OX14 4RN

Routledge is an imprint of the Taylor & Francis Group, an informa business

First edition published by Routledge 2021

ISBN: 978-1-032-76252-4 (hbk)
ISBN: 978-1-032-76250-0 (pbk)
ISBN: 978-1-003-47858-4 (ebk)

DOI: 10.4324/9781003478584

Typeset in Palatino
by codeMantra

Find additional online teacher resources, discussion boards, and other community resources at: www.EmpoweredScienceTeachers.com

For John, Westin, and Russell Lê.
Thanks for being my favorite everything.
Let's be where our feet are,
and remember what we're made of.

To teachers everywhere.
The world will soon recognize
that we are the critical mass needed
to fight the climate emergency.
It is both a superpower and a privilege.
We got this.

Contents

A Pulse on the Current Climate • A Moment to Reflect
• Beginning Where You Are • Shared Leadership
• Transparency • Overview of Chapters • Learning Icons
• Trust the Process

A Story about Tradition • An Opportunity to Challenge
Science Education • Unveiling Your Teacher Disposition
• Limitations with the Scientific Method • Diving Deeper
into the Process of Science • Getting Cozy with the Nature
of Science • Equity and Antiracist Science Teaching
• Putting the Pieces Together

Critical Perspectives for Complex Thinking • A Second
Look at the NGSS for Climate Science • A Fresh Take on the
NGSS • The Role of Climate Literacy & NGSS • Climate
Change as a Socioscientific Issue • Deeper Dive into SSI to
Inform Instruction • Moving Forward with Confidence

Meet the Author

Kelley T. Lê, EdD, is inaugral the Executive Director of the University of California and California State University Environmental and Climate Change Literacy Projects (UC-CSU ECCLPs). This initiative is the first of its kind to connect every public university and thousands of environmental community leaders across one state to advance PK-12 climate and environmental literacy, justice, and action for pre- and in-service teachers. Dr. Lê served the educational field for over 15 years in various roles. She was a high school chemistry/earth science/ and nanoscience teacher, instructional coach, induction mentor teacher, consultant for programs such as the UCLA Curtis Center and UCLA Science Project, coordinator for UCI CalTeach Math & Science, and program director for the UCI Science Project. Her experiences led her to also design new courses including a high school nanoscience course supported by UCLA C(n)SI that centered on climate change technology and innovation, as well as a university-level course for pre-service math and science candidates to learn about pedagogy using climate change investigations to anchor their experiences. Kelley T. Lê has supported thousands of educators in the areas of climate change, environmental justice, nanoscience, next-generation science, transformational coaching, and mentorship. She is also the proud recipient of the National Center for Science Education (NCSE), Friends of the Planet Award (2022) as a result of her ongoing advocacy efforts.

Dr. Lê graduated from Leuzinger High School in the inner city of Los Angeles, and is a First-Generation Southeast Asian social justice leader. She advocates for justice due to her lived experiences, but also received formal educational training from Loyola Marymount University, CSU Dominguez Hills, UCLA's

Center X teacher education program, and UCLA's Educational Leadership Program (ELP). What continues to influence her both as an environmental educator and a lifelong student is her love for dancing Hawaiian hula with her hula ohana.

List of Abbreviations

AP	Anchoring Phenomenon
CCC	Crosscutting Concepts
CCE	Climate Change Education
CER	Claim, Evidence, Reasoning
CRRP	Culturally Relevant and Responsive Pedagogy
CSU	California State University
DCI	Disciplinary Core Ideas
ECCLPs	Environmental and Climate Change Literacy Projects
ESS	Earth and Space Science
ETS	Engineering, Technology, and Application of Science
GCC	Global Climate Change
GHG	Greenhouse Gas
IP	Investigative Phenomenon
ISTE	International Society for Technology Education
MLL	Multilingual Language Learner
NAS	National Academies of Science
NASA	National Aeronautics and Space Administration
NCSE	National Center for Science Education
NGSS	Next Generation Science Standards
NOAA	National Oceanic and Atmospheric Administration
NOS	Nature of Science
NRC	National Research Council
NSTA	National Science Teachers Association
PL	Professional Learning
POS	Process of Science
SEL	Social and Emotional Learning
SEP	Science and Engineering Practices
SDG	Sustainable Development Goals
SSI	Socioscientific Issues

STEM	Science, Technology, Engineering, and Math
TFGCC	Teacher-Friendly Guide to Climate Change
UC	University of California
URM	Underrepresented Minority
US	United States
YPCCC	Yale Program on Climate Change Communication

List of Exhibits

List of Figures

List of Tables

Foreword

Frank Niepold, Senior Climate Education
and Workforce Program Manager
NOAA Climate Program Office, Communication,
Education and Engagement Division

On Monday October 30 in 2017 I received an interesting email from NOAA Public Affairs. It was from Kelley T. Lê, a science educator and department chairperson at the time for a Title 1 school in Los Angeles and was "extremely passionate about disseminating climate change education to as many students as possible." She was looking for and researching programs that support educators to teach climate change topics while pursuing her doctorate degree from UCLA. As the NOAA Climate Education Coordinator, I had been working to support teachers to incorporate the climate content into the classroom since 2005. Later that day I called Kelley T. Lê and we began to partner to help teachers use and leverage the NGSS to increase students' scientific literacy, skills, and knowledge to effectively address the challenges and opportunities that climate change presents to communities across the nation.

Over the years, Kelley T. Lê and I have explored this shared mission in numerous workshops, dialogs, and the 2019 Environmental and Climate Change Literacy Project and Summit (ECCLPS) held at UCLA. At the summit we explored with participants the alignment between the NGSS and climate change topics (using the California 2016 Science Framework) across the grade levels. In our breakout we explored how learning progresses in the NGSS related to climate change, and how to support all teachers in California. At the summit, Kelley T. Lê mentioned she was working on this book and I was excited to see where her deep experience and passion would take her.

Addressing the climate crisis requires transformation of our social systems, food systems, electricity production, land use, and transportation. It also requires transformation in how we teach our coming generations who need to prepare for a very different world. I think this book provides a much needed support to help teachers envision their teaching as they start their journey of transformation.

Since the original forward to this book, at the (re)launch of ECCLPs held on September 15, 2022 at a University of California—Irvine event Kelley T. Lê was announced as ECCLPs Executive Director. That event brought together key stakeholders to engage, support, and continue to advance new UC-CSU climate and environmental justice literacy initiatives. ECCLPs have launched a series of bold and transformational initiatives that are putting the work described in this book into practice.

Let me make the case for why this transformation of teaching is needed. Education not only enables society to benefit from the climate model projections. Climate science gives humanity choices of possible futures and a timeline to follow for each future. If we want a stable climate then we need to act together to choose the right path, a path that rapidly decarbonizes enough for life to thriving across the planet. These models show us the role of our collective decisions and natural systems that work together in reinforcing feedback loops. Our education systems need to accelerate the growth of climate-informed decision-making at all levels. It needs to build social will and support for action, and enable the climate-ready workforce. Education in all its diverse forms (science, STEM, environmental, conservation, civic, place-based, etc.) needs to broadly support communities as they transition to low-carbon and resilient economies. The 2009 Climate Literacy guide, which I led for the government, approved by 18 US federal science agencies and a consortium of science and education partners, concluded

> …as nations and the international community seek solutions to global climate change over the coming decades, a more comprehensive, interdisciplinary approach to climate literacy—one that includes economic and social

considerations—will play a vital role in knowledgeable planning, decision making, and governance.

(U.S. Global Change Research Program, 2009, p. 8)

Teachers across the country are preparing to teach students the science and engineering called for in the new science framework. These new concepts address global challenges and opportunities posed by climate change, such as generating sufficient clean energy, building climate resilience for businesses and communities, maintaining safe supplies of food and clean water, and solving the problems of global environmental change. Kelley T. Lê partnered with us in the US Global Change Research Program to hold listening sessions in 2023 to update the Climate Literacy Guide. We will release the new guide in the fall of 2024. We heard in these 21 listening sessions that the amount of time teachers spend on these issues needs to increase significantly and cover all disciplines and types of education. Education systems in some states and communities across the United States are developing and deploying new interdisciplinary models of education to support learners of all levels to foster climate, energy, literacy, and action. Armed with newfound knowledge and skills, students are increasingly able to contribute to and accelerate climate action in their communities.

Consider this updated book a call to action for you to join your fellow education leaders and teachers as we collectively transform our practices to prepare students for the climate solutions that are already underway. There are amazing and highly effective programs to support you and your students. Come explore them in this book and beyond, learn what works from others, and build your own agency and hope for the future so you can model this for your students and learners. Even as the science becomes more dire and the time to address this challenge shortens, today's students need to be ready to accelerate the climate solutions of today and tomorrow. So many impressive transitions are already underway, from solar and wind farms to reducing food waste and moving to plant-based diets. As Paul Hawken, lead author for the book *Drawdown*, said, "addressing it [climate change] is a pathway to transformation, creating a far better civilization than the one we live in now (Rysavy, 2017)." We need to inspire and support our students to design a better future that solves the climate change challenge while improving everyone's quality of life at the same time.

Today's youth are inheriting the unparalleled impacts of climate change. They are also among our most powerful champions for a sustainable, climate-resilient future. This elevates the importance of preparing today's students to implement policies and develop innovations needed to realize that future.

The education needs are significant, but they are magnified by a gap in hope when it comes to engagement on issues pertaining to climate change. Americans who have hope are more likely to engage in climate change solutions and to talk about it with their family and friends. Educators are critical messengers and facilitators of climate change education and have an opportunity to cultivate much needed hope for the future in their students. As educators, we need to know that hope is a precondition to action and these are times for action.

Over the last 18 years we have learned important things about how to close the education gap. Social science research is also clear: acquiring knowledge about climate change does not necessarily move individuals to action. Affective and social forces often influence risk perception and actions around climate change. Thus, knowledge must be paired with affect, beliefs, intentions, and motivation to enact change. The need for life-wide, comprehensive, and transdisciplinary climate change education is more important now than ever before.

We are privileged to live in a time when science offers meticulously observed climate trends and rigorously grounded climate model projections of possible outcomes, offering an opportunity to learn without requiring us to experience the outcomes our science sees as possible climate futures. What we do together today will determine our shared climate future. Yet this privilege cannot be realized without building that knowledge and skill through education. Let's do this transformational work together; it makes the heavy lift so much easier and more sustainable.

Foreword

Christina T. Kwauk

I grew up in an era where the specter of climate change had not yet permeated my nor my peers' consciousness. Of course, climate change existed *somewhere*: nestled within the environmental messages of children's books like *The Lorax*, subsumed by science lessons on the greenhouse effect or the last Ice Age, as the setting of futuristic science fiction novels like *Parable of the Sower*, or hidden in contemporary disaster movies like *Deep Impact* and *Armageddon*. Climate change was either something that marked geological periods of Earth's planetary history or as part of the subtle warning provided by creative visionaries of what *could be,* someday in the future, a harsher existence for humankind on the planet we call home. Climate change was the stuff of social commentary—or a scientific concept about the natural environment at school—not the existential threat that it is today.

So let's fast forward to today, where my own children are growing up in an era where hurricane season is more intense, where their father's favorite childhood beaches in Hawai'i have eroded beyond belief, where each year breaks another heat record, and where the smell and haze of distant wildfire smoke have all been normalized—events and phenomena that have lost their shock value and, perhaps, their ability to urge action. For my kids, climate change as a concept is no longer of the distant speculative future or of the distant geological past; it is now. As a concept, it is no longer a natural phenomenon absent of human influence; rather, it is fueled by the last three centuries of human activity.

Seeing climate change for what it is—an existential threat to humankind—demands a paradigm shift among our education system—and thus among us educators—when it comes to *what* we teach (from "just" the climate science to climate change as socioscientific issue), *how* we teach (from just the facts to problem-based learning, antiracist teaching, and storytelling), *when* we teach (from isolated topics in specific units to relational and interdependent phenomena across the curriculum), *where* we teach (from inside the classroom to our built and natural environments), *why* we teach (from fulfilling a curricular criteria to building competencies for climate action, climate resilience, climate justice, and a greener world of work), and to *whom* we teach (from children as recipients of knowledge to children with lived experiences of the impacts of climate change and related environmental injustices). My children, your children, our learners, and their futures can't afford for us grownups to continue "education as usual." Their wellbeing and the wellbeing of our future generations is dependent on us grownups wholly repurposing and reshaping education today for the Anthropocene.

With that framing, this second edition of Kelley T. Lê's book couldn't come at a more critical time for education in the United States and for science educators across the country. This book offers important frameworks and tools for reflection for science educators to collectively spark the educational paradigm shift required of our epoch. And what is special about this book is that it puts in the hands of every science educator an accessible and personal guide to navigate what can be an emotional journey of finding one's role, one's voice, and one's place in climate action.

We all have our own stories to tell about our climate journeys—that string of experiences that led us to recognize the extent of the climate crisis and the scale of solutions needed, and when layered with our love and care for the next generation fed in us a desire to take action, personally and professionally. These stories are many. And each story offers a glimpse into an array of nonlinear, imperfect, serendipitous entry points to climate action.

My own story is one of these. I am not your typical feminist climate activist. On the contrary, as a child of immigrants from China, being an activist of any nature was looked upon with great suspicion. We were a politically conservative household assimilating to America in our home in northern Alabama. I viewed the outdoors as a place that was wrought with dangers:

relentless mosquitoes that seemed to only want to suck *my* blood; Red Paper Wasps that lived under our porch and threatened to sting me anytime I left the house; ladybugs that infested my room every spring; poison ivy wherever I spotted leaves of three. If the outdoors were to be enjoyed, it was places like Niagara Falls or the Great Smokey Mountains. Places you drove to by car, wowed at the beauty, smiled for the camera, and then afterwards returned to the safety of the air conditioned indoors.

While I didn't grow up a feminist climate activist, I did grow up with a very acute sense of power inequalities between people and how these created unfair advantages and disadvantages for certain populations. And it was that perception, nurtured by the tools of ethnographic analysis, critical discourse analysis, and policy analysis that strengthened my ability to connect the dots between seemingly disconnected issues, like gender equality, transformational learning, and climate justice. And one day, during a summer work trip to London when I was in a meeting with a former world leader to discuss a new girls' education initiative we were spearheading, I stumbled upon my entry point to my climate journey. This world leader, eating her breakfast across the table from me, asked me a completely innocuous question: What is the impact of climate change on girls' education (in low and middle income countries where gender inequality in education is high)? I had no idea! I wasn't a climate scientist! But surely, knowing that climate change was a "bad" thing meant that it would threaten the progress girls' education activists have made over the last few decades? It was that moment, that question, that led me down a diverse set of paths of inquiry and action to where I am today: a social scientist trained in gender, education, and international development working now to help move the education sector toward climate action and the achievement of climate justice.

Not your typical origin story for someone working on issues of climate change! But the point is—and the underlying assumption of this book is: there is no typical origin story of "the climate activist" or a typical entry point to becoming "the climate change educator." The times—and our children—are demanding that we *all* become climate activists and climate change educators.

Our individual climate journeys—your climate journey and your foray into this book—are important pieces of a larger tapestry that will tell the collective story of humanity's diverse, imperfect, challenging, yet nonetheless necessary response to a warming planet. What will be your moment, your origin story to becoming a climate change educator?

Acknowledgments

In 2024, I came across a video of basketball player Jeremy Lin describing what led to his "last game" with the New York Knicks. It's important to know that leading up to his breakout game, Lin's assistant coach advised him to "not to do too much," which resulted in him playing it safe and underperforming. Afterwards, Lin's agent called to inform him that the next game would be his last and that he might as well go out giving it everything he's got. His exact words were, "If you get put in the game, you gotta play Jeremy Lin basketball." That was when the world witnessed *Linsanity*! His message to the audience during this moment of reflection? **Think about how you would show up if you were uninhibited—as though it's your last opportunity to make a lasting impact.**

In reflecting on the work of centering on climate change education as the most powerful climate action we can take (or at least support) because it catalyzes the necessary innovation and solutions needed, I find myself asking the following:

What would I do if I were uninhibited?
How would I do those things?
<u>*Who am I doing this for?*</u>
Can I start today?

Of all the questions, the third one is the one that motivates and inspires me to keep showing up. First, I'm doing this for my family. I'm reminded constantly that I come from a long line of strong, resilient, and resourceful survivors who fought for democracy in Viet Nam, and survived the unimaginably treacherous journey as refugees of the war. Facing incredible discrimination, racism, and injustice, their strengths lie within me and my children. To my mom, Jackie Ho, thank you for the sacrifices you made to give us the life we have today. Never one to back down from injustice, I know why fighting for equity and justice are both a responsibility and privilege because of how you raised me. To my high school sweetheart, John Lê, who made my name rhyme and more confusing in spelling with all the extra "L's" and the "E's," one lifetime together isn't long enough. Thank you for being my favorite person in all my past lives and in the future ones. For Westin and Russell, thanks for being all the best parts of me, your dad, your grandparents, and all your ancestors before. Each generation is lifted a little higher than the last, I hope that when you're older you look around and see a greener, more just world that we are working hard to build. Stand firmly in your beliefs and values so that you don't go so easily in whatever direction the wind blows.

For my former students who pushed, challenged, and inspired me over the years, thank you for letting me be part of your journey even if it was a small moment in your life. I hope that when people ask you to reflect on your high school science days, you remember it with great joy in a time where we could dream of a more just community and society together. Thank you for being a central part of my leadership journey and advocacy efforts.

A huge thank you to all my colleagues who have worked alongside me and elevated my voice. I thank you for being co-conspirators and allies to help me navigate. You know who you are! I especially would like to give a shout out to partners sprinkled throughout this second edition. You personally inspire and uplift me and my climate change education efforts. Thank you for being incredible humans.

Finally, thank you to Julia Dolinger and the Routledge team for giving me the opportunity to reach more teachers and students through the second edition of this book. Your support was instrumental to bringing this to life.

Introduction

This book was written to support science educators looking to create transformative student learning experiences centered on climate change by leveraging the Next Generation Science Standards (NGSS) Framework. As a former teacher and instructional coach, I personally navigated many challenges and successes that came with taking on climate change education and NGSS. As a current educational leader, I'm going to highlight all that I have learned because sharing is caring and I won't be ionic. This guide will help you (1) understand your teacher identity and your pedagogical decisions (because there are reasons why even the best professional development (PD) doesn't transfer into the classroom sometimes), (2) provide you with realistic ways to take on both climate education and NGSS (because climate and the environment are perfect vehicles to engage students while addressing the framework), and (3) build your capacity to activate students through this work (because we recognize the deep yearning for climate action and justice that can no longer wait). We know that climate change impacts everyone, but what is often overlooked are the social, racial, and environmental injustices that are further exacerbated by this crisis making the impacts disproportionate. Given that the majority of people rely on mass media for information and education on climate change (Caranto & Pitpitunge, 2015; Carter & Wiles, 2014; Hestness, McDonald,

DOI: 10.4324/9781003478584-1

Breslyn, McGinnis, & Mouza, 2014), it is crucial for teachers to learn key information and facts along with research-based teaching practices that lead to supporting students as informed and capable community agents of change.

A Pulse on the Current Climate

In a report published by the National Public Radio (NPR), 80% of parents support the teaching of global climate change in America, but 55% of teachers say they don't teach about climate change because it doesn't relate to their content areas (Kamenetz, 2019). Prior research also reveals that teachers who do teach about climate change, only spend between 1–2 hours on the topic each year (Plutzer, McCaffrey, Hannah, Rosenau, Berbeco, & Reid 2016). Furthermore, there are major teaching inconsistencies due to a myriad of factors including the need for deep climate science content knowledge (because teachers don't know what they don't know), scientific literacy (because climate science was not likely built into our own training programs and we cannot keep relying on mass media), or access to evidence-driven curriculum (Bunten & Dawson, 2014; Dawson, 2012; Hansen, 2010).

In a survey of over 2,000 high school students conducted by the Environmental and Climate Change Literacy Projects (ECCLPs) in early 2024, findings show that over 90% of students believe that climate is happening now, but their concerns are out-sized by their knowledge in being able to explain the causes and consequences of the emergency. The top three sources of information cited by students in looking at where they get their climate education includes social media (74%), news (62%), and friends (37%). Also noteworthy, students are very comfortable in their abilities to make a positive environmental impact underscoring the important finding that **students who take at least one action, stay taking action to fight climate change as a source of hope.** Knowing what motivates students and what they want to learn more about can help shape what we decide to teach in the classroom to further engage and activate them.

That's why **education <u>for</u> climate action** is critical for teachers taking on this politically (but not scientifically) controversial topic to help students become informed decision-makers. It is essential to teach today's youth about the climate crisis that is quickly changing the world they are inheriting. As teachers, we hold tremendous power in the classroom to decide what is of value in that space. So how do we integrate climate science and environmental topics? How might we use climate change to engage students in cyclical and iterative ways of thinking to bring deeper meaning to science education? How might we move from messages of "Gloom and Doom," to helping students access their own agency as leaders who will mitigate the devastating impacts? This book will address those common questions and provide implementation support to take it back to the class. But first, let's meet you where you are. We'll begin with what you know, are currently great at, and how you envision taking action through education based on your spheres of influence.

A Moment to Reflect

Let's start this learning journey by first identifying your intentions as well as relevant learning and implementation goals. The reasons why teachers want to teach about climate change vary greatly, and identifying your lens is an important first step to naming your teaching disposition (e.g. attitudes, beliefs, behaviors, and characteristics you approach teaching with). Take a moment to consider these questions to get started: Are you purely looking for content and/or teaching support? Are you looking for ways to empower students to take action? Are you looking for ways to engage students as critical thinkers? Are you wanting to co-construct culturally relevant and responsive science experiences with students?

I invite you to think about reasons for why you are interested in taking on climate change, and what you're currently wondering about in regard to teaching it. Then, I encourage you to note possible expectations and goals you might have for this book, to hold yourself accountable to bringing new ideas back to your class. Finally, identify realistic challenges that may prevent you from making or implementing new ideas.

I'm going to level with you. Of all the PD that I have attended (by choice or by contract), there were likely only a handful that were so "window shattering" that I could not wait to enact some version of it with my own students. We know that what teachers often learn in PD rarely makes it back to the class and that's also known as "The Problem of Enactment." The Problem of Enactment stems from teachers not implementing new strategies, tools, or learnings because it likely doesn't align with their personal teaching beliefs and values. This can also happen when teachers don't feel the facilitator is relatable (completely subjective on this one). As a result, educators tend to revert back to teaching the way they personally experienced, were trained to do, or what has worked in the past. It is important to remember that teachers are individuals who have successfully navigated the current schooling system. Also critical, is understanding that this system was never designed with the intention of supporting all students or to serve as organizations for societal change. In fact, many argue it exists to perpetuate the status quo, which is why a paradigm shift is needed to reposition the role of schools which can start with us. Consider your teacher disposition as you review the following questions and note what you are ready to explore.

- ♦ What evidence reveals that the current educational system is working (or not) to support all students?
- ♦ How might climate change disrupt schools and how schools function? What about how and what we teach?
- ♦ What are the different ways of learning (beyond what is passed on through formal schooling) climate change, the environment, sustainability, resilience, or solutions? Who are the teachers? How are students positioned?
- ♦ What shifts need to take place to ensure that the current educational system will support the next generation of environmental stewards and leaders needed to mitigate or adapt to the climate crisis?

Continue to Exhibit A to help you set clear intentions and implementation goals based on your reflections. As you explore the chapters, think about what risks you are willing to take (or challenges you might encounter) as you learn ways to transform your teaching practices and curriculum.

Exhibit A: Setting Clear Intentions and Learning Goals Activity

1. Some of the reasons for why I want to teach about climate change are…
2. I expect this book to help me…
3. My realistic learning goals prior to digging into this book are…
4. Right now, a possible challenge to bringing these ideas back to my class are…
5. Possible fears or concerns I currently have in regard to teaching about climate change are…
6. I recognize potential "risks" that might prevent me from taking this back to the class include…

   a. **Risks related to my** *personal* **growth** (might be taking a risk by teaching new content, feeling imposter syndrome, not having enough time to plan, not familiar yet with teaching ethical dimensions of science topics, burnout, etc.).
   b. **Risks related to my** *professional* **growth** (might be taking a risk by having to teach this alone, trying unfamiliar resources, not having enough time to teach it, not having site support, teaching a politically controversial topic, etc.).

There are many activities layered throughout this book that provide crucial opportunities for reflection and application of knowledge. The first step to successfully learning new material starts by unveiling our current understandings (what we currently know) and our Funds of Knowledge (How we came to know that and what cultural influences shape that understanding).

Beginning Where You Are

As teachers, we all hold beliefs. We have beliefs about our jobs, roles, colleagues, students, school, and more. A belief is an idea or assumption that we have accepted to be true. The thing about beliefs that we often forget, however, is that beliefs *can* change over time. As you explore this guide, you might find yourself questioning some things you believe about educational institutions, climate change, or the NGSS. As you begin to develop different ways of thinking about climate science, it is important to reflect on your teaching practices and examine what would be worth bringing back to the class for students. The next activity will gauge your current beliefs around student learning experiences, curriculum design elements, and teacher attributes.

Tracking Your Professional Progress

Student Learning Experiences in Your Class	On a scale of 1 (extremely disagree) to 5 (extremely agree), to what extent do I agree or disagree with the following statements? Why?
Today's Date:	
Students often discuss policies related to science.	
Students often collect and analyze current data or information.	
Students often discuss ethical issues related to science.	
Students often co-construct knowledge with me during lessons.	
Students often drive the instruction in my class as capable contributors and doers of science.	
Students are often positioned as current scientists or engineers.	
Students often learn about Indigenous traditional ecological knowledge.	

Curriculum Design Elements	On a scale of 1 (extremely disagree) to 5 (extremely agree), to what extent do I agree or disagree with the following statements? Why?
Today's Date:	
I often build lessons/units around anchoring or investigative phenomena.	
I often present climate or environmental issues at the start of each unit or lesson.	
Students often engage in argumentation and making claims based on evidence.	
Students often engage in meaningful discourse opportunities.	
My lessons/units are often centered around real-world issues that are directly related to students' lives or community.	
Students often engage with the Nature of Science principles.	
Students often connect with local or relevant field experts and researchers.	

My Teacher Attributes	On a scale of 1 (extremely disagree) to 5 (extremely agree), to what extent do I agree or disagree with the following statements? Why?
Today's Date:	
I have to know everything about a particular topic before teaching it.	
I am extremely confident in my knowledge of climate science.	
I feel comfortable admitting to students when I don't know the answer to their question.	

My Teacher Attributes	On a scale of 1 (extremely disagree) to 5 (extremely agree), to what extent do I agree or disagree with the following statements? Why?
I am comfortable teaching about open-ended issues where I cannot predict student responses.	
I am not the only source of knowledge on climate and the environment for students.	
I often experience imposter syndrome when teaching, even for topics that I have strong expertise in.	
I often reflect on my pedagogical decisions to improve my practice.	

We will revisit the following activity several times to help you track your professional growth and note any driving questions that emerge in the process.

Shared Leadership

As you may have already discovered, the topic of climate change is complex, emotional, and challenging. It is complex because it requires a systems-thinking approach to fully understand the scope of the problem, but teachers might organize and scaffold information in ways that make the topic seem segmented or disconnected. Climate change is highly emotional because it impacts all living things on Earth, and activating student agency requires teachers to also address the ethical dimensions (such as social, racial, and environmental injustices). Lastly, the topic is seemingly challenging due to its politically (not scientifically) controversial nature where people rely heavily on social media for information. To address these elements, teachers need to understand the Nature and Process of Science (also embedded in the NGSS) to develop strong scientific literacy in students.

> Better to light a candle than curse the darkness. There's so many people who sit back and say we're screwed, but you know what, with that one candle maybe someone else with a candle will find you, and I think that's where movements are started.
>
> – Shawn Henrichs

Consider the importance of a shared leadership approach at your school and what is needed to make that successful. In *Racing Extinction*, a documentary showcasing human impacts on species worldwide, the filmmakers remind us that we are truly stronger and will go further if we mobilize on these challenges together (2015). Consider collaborating with colleagues who are open and willing to design and implement with to show proof of concept. Remember the lens that you are looking through as you approach this book, and the fact that other teachers likely have different perspectives because of their own Funds of Knowledge and current understandings. By unveiling your colleagues' teacher disposition, you'll be able to have deeper conversations about curricular design and implementation grounded in deeply rooted values and beliefs that are shared. At the very least, we know we need to fulfill on three-dimensional learning for NGSS, and climate science makes up nearly 30% of the framework. Think about how you might leverage the standards to bring your colleagues together to teach about climate change and the environment. Start by reflecting on the following:

◆ Who are the teachers in other subject areas that might have a shared vision and belief that schools can support students to be informed decision-makers on current issues and challenges of today.

♦ Where might you find allies to help you navigate the bureaucracy of schooling that might prevent/slow related educational efforts? Who's currently teaching this? Leading in the community? Started a school club? Who are the students involved? What's the interest level?

We need to continue believing in our students, ourselves, our colleagues, and realize that we have power and privileges within our own spheres of influence to bend the curve for future generations through education.

Transparency

This book was written to provide teachers with accessible ways to integrate climate science, build their confidence around the NGSS, and provide ways to activate student agency. On the surface, this is a guide to teaching climate change while fulfilling the NGSS. Underneath, I hope to push teachers' thinking about how science has been historically taught and how the NGSS can catalyze climate action through meaningful learning opportunities for all. We are helping to raise the next generation of planetary stewards who will need to be resilient, resourceful, and hopeful about the future that awaits them. We have the power and privilege (acknowledging that some might have paths with more/less resistance than others) of teaching about climate change to give students every opportunity to reimagine and realize a better and more just world.

How we do this is up to us, and this book is meant to spark creativity in both design and communication of climate and the evolving environment. Take a look at the image on the book cover and answer the following questions:

1. *What does the image make you think of?*
2. *How does the image make you feel?*
3. *In what ways (if any) might you relate to the image that you see? Say more about that.*

4. *Is there anything in particular that stands out to you about the image that relates to climate change? Share more.*

If you struggled to find a connection to meaningful respond to any of the questions above, you're right where you need to be. Let's put a pin in this until after I share more.

The cover of my book was intentionally designed to showcase opportunities and spotlights that we tend to glaze over when thinking about climate and the environment. This is a classic example of how context brings meaning to the content to ensure that the "dog is wagging the tail" rather than the "tail wagging the dog" (Brown, 2019). The image shows various development stages of Robusta coffee beans that are growing in the fields of Viet Nam. Once considered the more undesirable plant compared to the more popular Arabica bean, Robusta is quickly rising as a top crop due to its ability to grow rapidly and robustly in challenging conditions. Viet Nam supplies nearly half of the world's supply, with crops that are more climate resilient with higher yields than anywhere else on the planet! As climate change continues to create more hostile environments, finding solutions to everyday problems will become a necessity and the new norm to ensure our survival.

To me, Robusta beans look like the climate stripes and the varying state of changes between the layers (view yours at *showyourstripes.info*). I think of nature-based solutions to bend the curve, and wonder about key approaches in relation to cultural significance and relevance. The clusters remind me of the importance of communities and how each is so different—yet I can also find similarities. I also wonder about the extraction of knowledge, resources, and demands on people's health. Specifically, about how instead of acknowledging and honoring these Indigenous practices that allowed for the beans to flourish in tropical climates, that erasure will take place and we'll force ecosystem changes rather than leveraging each other's strengths and collaborating so we can all *do more than survive* (Love, 2019). This parallels so much with education. Specifically, that reframing and respectfully unveiling hidden opportunities are critical as we prepare for what's ahead in a changing climate. *When will we collectively understand and see that education is both a hidden superpower and solution for catalyzing climate action? Like the Robusta beans, will there be a specific moment that turns the tide to finally reposition education at the forefront of bending the curve?* **The answer in short is yes and that moment in time is right now starting with us.**

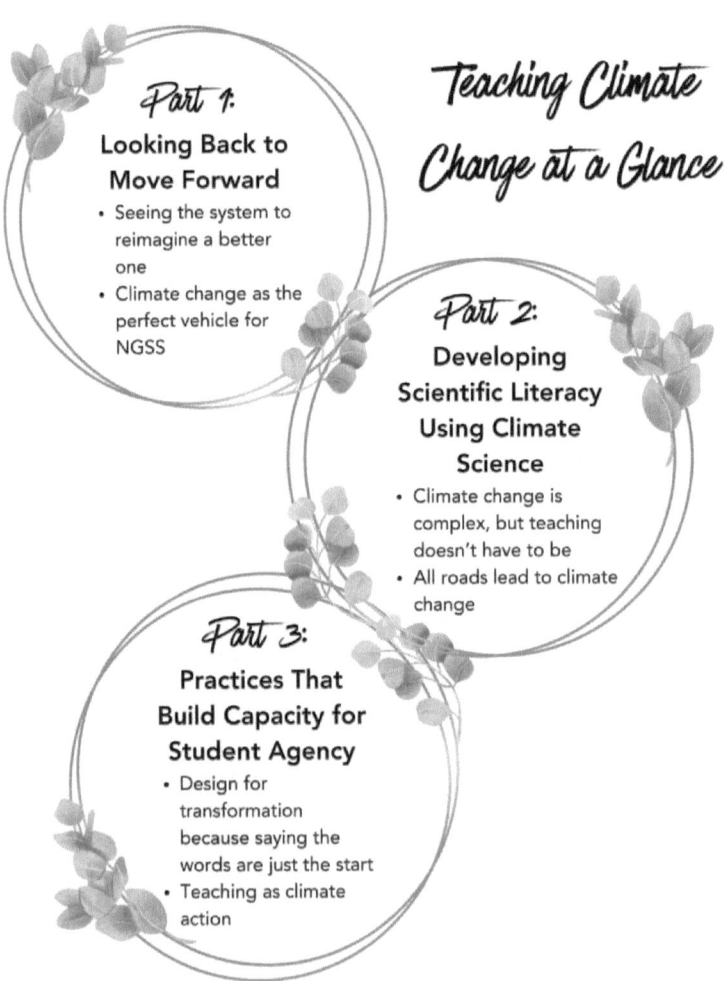

FIGURE 0.1 Teaching Climate Change at a Glance?

Take a moment to revisit the initial questions I posed. Has providing you deeper context and unveiling my own thinking, changed or affirmed any of your own answers? Next, you'll see the overview of chapters that will give you a glimpse of more reflection points ahead.

1. *What does the image make you think of now?*
2. *How does the image make you feel now?*
3. *In what ways (if any) might you relate to the image that you see now?*
4. *Is there anything in particular that stands out to you about the image that relates to climate change now?*

Overview of Chapters

Chapter 1, "Reimagining Science Teaching" is useful for teachers who are ready to examine their pedagogical practices through the lens of NGSS. Continuing from where you are, this chapter pushes you to reflect on how much of a paradigm shift you are ready to engage in as you take on climate science for transformational teaching.

Chapter 2, "Leveraging the NGSS for Climate Education" examines NGSS-aligned instruction and the crucial role of climate change in the framework. This chapter also provides a comprehensive overview of research and data behind the frameworks grounding this book, and how these approaches have proven to be effective for both student and teacher learning. There will be many opportunities to transfer what you are learning into the classroom to experience what a climate change NGSS-aligned lesson sounds, looks, and feels like with students.

Chapter 3, "Climate Change is Complex, Where do I Start?" goes over the fundamentals of climate science as outlined by national and local educational directors from various reputable organizations. This chapter also highlights what researchers in the field of climate science education have found to be effective in teaching about climate change both in science and beyond the field.

Chapter 4, "Climate Change as the Anchor" explores what it looks like to center science instruction around climate science to empower students as agents of change. Using climate and environmental phenomena as anchors in NGSS storylines, teachers receive practical tools and resources to develop students as complex critical consumers and thinkers.

Chapter 5, "Planning and Teaching for Transformation" provides a comprehensive look at lesson planning NGSS storylines, while learning about paradigm shifts that need to take place to effectively teach climate change to activate agency. Teachers will also explore the ethical dimensions of climate change with students through culturally relevant and responsive approaches.

Chapter 6, "Education as the Catalyst for Climate Action" provides additional support for teachers looking to transform their practices and pedagogy as equity and justice-centered educators. This final chapter explores a myriad of teaching resources aligned to NGSS, while further tackling the ethical dimensions of the climate crisis to incorporate ways for students to be change agents in their own communities.

Frameworks for Successful Teaching & Learning

There are three frameworks referenced often throughout the book to support the teaching and learning of climate change content through NGSS. See Table 0.1 for the frameworks and their descriptions.

The appendices at the end of the book include referenced tools, resources, and samples of work. Visit https://www. empoweredscienceteachers.com/ to find out more about my current leadership efforts, access teacher resources, and get information on current professional learning opportunities.

TABLE 0.1 Major Frameworks & Descriptions

Framework	Description
Culturally Relevant & Responsive Pedagogy (CRRP)	This framework recognizes the value of students' cultural backgrounds, Funds of Knowledge, and lived experiences in education. Ladson-Billings (1995) stresses the need for educators to tend to student achievement and learning, cultural competence, and sociopolitical awareness/critical consciousness to empower students.
How Students Learn Science	Published by the NRC (2015), researchers take into account how students learn science successfully in K-12 schools. This body of research emphasizes three student learning principles.
Next Generation Science Standards (NGSS)	Released in 2013, this science framework includes three-dimensional learning and mastery of performance expectations. There is also an emphasis on engineering practices, climate change content, and the Nature of Science.
Socioscientific Issues Framework	This framework focuses on science topics that are complicated, open-ended, and controversial without straightforward solutions (such as climate change). It also looks at ways to address climate change effectively for both teachers and students and acknowledges the need to address relevant ethical dimensions.

Learning Icons

Throughout the book you will see the following icons displayed at specific points. These icons highlight opportunities to understand (1) where you are in your own practices, (2) when to connect with the network, and (3) ideal moments to try out what you learned or created with your classes.

(1) Learn More
About Myself

(2) Connect with
the Network

(3) Take it Back to
the Class

Trust the Process

"Choose courage over comfort" and remember that change takes time. It is essential to have small ambitious goals as you take on climate change in the classroom, but remember to have patience for yourself and to consistently return back to those goals to reflect on your progress. The NGSS pushes teachers to model the Nature and Process of Science for students, and although climate change allows for teachers to lean into this more easily, the topic has such a large scale and is so complex that it is impossible to know everything about it. Thank you for trusting the process as you navigate ways to effectively teach about climate change to activate student agency. Let's do this!

References

Brown, B. A. (2021). *Science in the city: Culturally relevant STEM education*. Harvard Education Press. https://hep.gse.harvard.edu/9781682533741/science-in-the-city/

Bunten, R., & Dawson, V. (2014). Teaching climate change science in senior secondary school: Issues, barriers and opportunities. *Teaching Science*, 60(1), 10.

Caranto, B. F., & Pitpitunge, A. D. (2015). Students' knowledge on climate change: Implications on interdisciplinary learning. In *Biology education and research in a changing planet* (pp. 21–30). Singapore: Springer. https://www.researchgate.net/publication/299940251_Students'_Knowledge_on_Climate_Change_Implications_on_Interdisciplinary_Learning

Carter, B. E., & Wiles, J. R. (2014). Scientific consensus and social controversy: Exploring relationships between students' conceptions of the nature of science, biological evolution, and global climate change. *Evolution: Education and Outreach*, 7(1), 6.

Dawson, V. (2012). Science teachers' perspectives about climate change. *Teaching Science: The Journal of the Australian Science Teachers Association*, 58(3) pages 8–13.

ECCLPs. (2024). *ECCLPs HS student survey report*. ECCLPs. https://ecclps.net/hs-ss-report

Hansen, P. J. K. (2010). Knowledge about the greenhouse effect and the effects of the Ozone Layer among Norwegian pupils finishing compulsory education in 1989, 1993, and 2005—What now? *International Journal of Science Education*, 32(3), 397–419.

Hestness, E., McDonald, R. C., Breslyn, W., McGinnis, J. R., & Mouza, C. (2014). Science teacher professional development in climate change education informed by the next generation science standards. *Journal of Geoscience Education*, 62(3), 319–329.

Kamenetz, A. (2019, April 22). *Most teachers don't teach climate change; 4 in 5 parents wish they did*. Retrieved May 25, 2020, from https://www.npr.org/2019/04/22/714262267/most-teachers-dont-teach-climate-change-4-in-5-parents-wish-they-did

Ladson-Billings, G. (1995). Toward a theory of culturally relevant pedagogy. *American Educational Research Journal*, 32(3), 465–491. https://journals.sagepub.com/doi/10.3102/00028312032003465

Love, B. L. (2019). *We want to do more than survive: Abolitionist teaching and the pursuit of educational freedom*. Beacon press.

Plutzer, E., McCaffrey, M., Hannah, A. L., Rosenau, J., Berbeco, M., & Reid, A. H. (2016). Climate confusion among US teachers. *Science*, 351(6274), 664–665.

Part 1

Looking Back to Move Forward

Reimagining Science Teaching

Read this when:

- ♦ *You're ready to think about your own pedagogical practices (ways of teaching) to determine where you are as we align to the NGSS.*
- ♦ *You're ready to engage in a paradigm shift (ways of thinking and knowing) to intentionally transform your practices.*

A Story about Tradition

Let me start with a story about a traditional practice that is still widely shared across many Asian cultures today. As a First-Generation Vietnamese-American, my family impressed upon me the importance of shared meals at the end of each day. During one of our family dinners, we began prepping the table for a pretty common Vietnamese dish known as Cá Nướng (or whole-baked and scallion oil marinated catfish). When each took their seat, it was customary to "mời (invite)" each person to eat starting with the eldest at the table to the youngest as a form of respect. I don't know when these rules were shared with me, but they felt culturally ingrained. My cousins and I were taught that when eating any type of whole fish, you are *never* allowed to flip the fish over to get the meat on the bottom side. It is simply

DOI: 10.4324/9781003478584-3

bad luck for you and your entire family. Many of us grew up with this omen and held it to be a sacred truth (for deep fear of being scolded and looked down on by the elders even when they weren't around). The older we got, however, the more questions we had about these traditional practices. The "window shattering" moment for me came during another one of those family dinners where Cá Nướng was served again. Randomly, my older cousin decided to ask his mother to explain *why* it's bad luck to flip the fish. Without any hesitation she explained that if you flipped over the fish, your boat would also meet the same fate and flip over in the ocean.

As a people who sustained and nourished themselves through sustainable agricultural and fishing practices, whom many were driven to flee by boat in the night to escape the war only to be at the mercy of what awaits them in the dangerous and treacherous journey across the ocean, it made a lot of sense to not tip the scales by doing anything that would increase your bad luck. After hearing this answer from my aunt however, we all burst into laughter after my cousin respectfully replied, "But we don't even have a boat." That question led us to abandon that practice as a family, while simultaneously holding a deep sense of gratitude for our elders who did whatever it took to survive.

 *Is there a similar story about a time-honored tradition that you decided to break away from once you learned more? Take a moment to think and jot down some thoughts in the margins.*

When you think about this story, it parallels how people still think about education and teaching today. Consider how much (or little) education has evolved over the decades compared to technology, communication, or nature. Although there is a vast amount of research detailing how students effectively learn, most of those findings rarely make it into the classroom. I encourage you to consider how some components of teaching fall in line with historical educational traditions (i.e. practices used to train factory workers, maintain the status quo, force assimilation, etc.), and compare it to research-based teaching practices on how students learn. When you think about aligning your practices and curriculum to the NGSS, it's important to identify what can and cannot be changed about teaching. You might not have the power to change the overall structure of education *yet* (i.e. bell schedule, instructional minutes, academic calendar, student-teacher ratio, etc.), but you do have the power to change your beliefs about what should be taught and how. It begins by asking the right questions about teaching and learning, and then doing your due diligence to see what credible researchers have discovered about your wonderings. I invite you to complete Exhibit 1.1 to start unveiling your underlying beliefs about teaching and learning.

 *As you begin to think about your current teaching practices in relation to climate science and NGSS, start by identifying realistic learning goals along with actionable steps. Write down your goals to see if they change over time.*

Exhibit 1.1 Reflective Questions to Recenter Your Teaching

Select one of the following questions that you want to personally explore more. Challenge yourself to think about how your answer reflects deeply held beliefs about the purpose of schooling. Do your answers affirm your thinking and/or do you have more questions to explore?

♦ In what ways has schooling *stayed the same* since you were a student in the PK-12 system (ex. are there practices used with students now that you remember experiencing when you were their age)? Why do you think society has continued to uphold those elements of schooling? Who might uphold or disrupt these traditions?

♦ In what ways has schooling *changed* since you were a student in the PK-12 system? What about changes in the last century? Why do you think society has pushed to change those elements of schooling?

◆ What have researchers uncovered about how students effectively learn science? What information might be useful to teachers? What might be missing from this field of research preventing classroom implementation?

◆ If your class was canceled today and never to be taught again, what would students be missing out on that they couldn't learn from ChatGPT or the internet?

◆ How might power dynamics between teacher and student impact classroom learning? Does that reflect in our lesson planning?

◆ How might students be authentically positioned as co-constructors of meaningful learning experiences? Who holds knowledge in the classroom? Is that reflected in our lesson design?

◆ Who decides what and/or how educational content should be taught? When are we in our power to decide? When are we forced to compromise?

◆ Is there a need for culturally relevant or responsive teaching? How might these types of learning experiences support students? Which students? In what ways?

◆ When do you tend to equity-centered instruction? What does that look, sound, and feel like in your class?

 The questions above introduce key ideas that support essential paradigm shifts needed to address 21st century challenges such as climate change. Beyond introducing short term strategies to integrate resources (which the internet has no shortage of), this book pushes you to reflect on your teaching disposition to support you in adopting or adapting curricular resources grounded in firm beliefs and values. When professional development fails to transfer to the classroom, there are deeper reasons. So let's start at the root with your expectations of good teaching and learning, where those ingrained traditions stem from, and why they matter.

An Opportunity to Challenge Science Education

Don Duggan Haas, the Director of Teacher Programs at the Paleontology Research Institute (PRI), affirms on their Science in Virtual Pub series that "the most valuable 'things' in our educational system are the human resources, especially the wisdom and passion of the teaching force (2020)." Haas (2020) argues that every part of society has evolved including the technology and innovations for entertainment, energy capturing systems, and transportation systems among others. Surprisingly, the one system that has not seen changes in the last century or so is education. It's safe to say that the schooling system has largely remained the same in terms of its design, built environment, feel, and functions because it wasn't initially designed to meet the needs of all students (nor did it carry the intention of positioning itself as an organization of community or social change). But we know it *can* change because systems are created and reinforced by people, and that means that our daily decisions in the classroom can uproot and challenge the system if we're intentional with our curriculum design and delivery.

Haas (2020) also notes that real problems that people care about are highly intersectional, such as the climate crisis happening all around us. What is needed to mitigate the impacts will be a transformation of how we think, feel, and act in our daily lives. That is not an easy ask. He also shares that people tend to adopt innovations, "When they are different enough to make a difference, but not too different that you don't understand it (2020)." Essentially, teachers tend to adopt new practices that are not too far off from what they have already been doing that can yield better results. I would re-emphasize here that teachers also tend to adopt new practices that are aligned with their underlying core values of what they believe reflects good teaching and learning. It might be in these quiet moments of reflection, that your "window shattering" moments occur.

I have attended multiple PDs where speakers revealed that only 10% of what we *hear* is actually retained, and that people generally remember more of what they can *see* or how you made them *feel*. When I am asking you to reimagine science education, it can be difficult to engage when you can't "see" the complex system. This makes it more difficult to understand our roles and how we contribute to and/or actively work to improve that system. Being able to visualize the system is an essential first step to understanding how our teaching reaffirms and/or challenges it. Here's what we know about the educational institution. There are "actors' or people that make up the institution who have various roles to keep it functioning. Take a second to visualize those people. We can also visualize good teaching when we see, hear, and feel it as we think about different classes in session throughout the day. It's interesting that many of us can also easily point out schools across communities (because they generally have an institutionalized "look" and "feel"). Consider what makes a school a school, why that is, as well as who makes those decisions and what beliefs they might hold based on what we see (or not) today. Complete the activity in Exhibit 1.2 to continue uncovering the layers of schooling and our roles inside of it.

Exhibit 1.2 Seeing the House Activity

How can we reimagine and/or rebuild an entire system when we're part of it and in different ways we directly contribute to it? Take a moment to engage in the following activity to help you visualize the schooling system.

Imagine that you're walking down a dirt pathway lined with grass and flowers on both sides. The sky is blue, there are fluffy white clouds, the Sun is high, and you can hear birds in the distance while a breeze grazes your shoulders. As you keep walking down this path, you find yourself walking toward a house a few steps away. *This house will symbolize the schooling system.* You first notice the colors of the house, the textures of the materials that make up the house, and a front door that is closed with one keyhole. You are given a key to enter this house, and you instinctively place it in the keyhole to see if it will open the front door. The door opens for you.

◆ What do you see when you look around?
◆ What is functioning well? What isn't?
◆ What are people doing inside this house? What are their roles?
◆ How was this house constructed?
◆ You notice different rooms in this house, what do those rooms represent?
◆ What's buried underneath the rugs (the unspoken things)?
◆ If values represented the foundation of this home (symbolizing the schooling system), what would those values be? Whose values would be represented?
◆ If *your* values are represented in the foundation of the house, what would those values be? In what ways are they similar or different from the ones you listed previously?
◆ If you are part of this home, what might be your role?

Pause here and reflect on the responses you provided to the questions above. To help you think about your direct role in this home that represents the schooling system, considering answering the following:

◆ Whose home is this?
◆ Who built it and why was it created?
◆ What are the rules of this house and who makes or enforces them? Who can change them?
◆ Who might have given you the key to enter this home? Did you have a hard time opening the door? Does everyone have the same experience with the keys they were given while trying to open the door?

Take a breath. Bring yourself back to where your feet currently are as you read this. Remember that the house symbolizes the schooling system that you are currently part of. **Think about the questions above and recenter on your teacher dispositions, beliefs, and values about**

teaching and learning. To dismantle a system, you first need to be able to grasp the scale of change needed and think about your active role in that system. Systems are created by people, which means that people within and beyond can change them. What levers of change will you need to pull on to understand the functionality and components of this system, redesigning any of the spaces within your control, or actively dismantling the system and rebuilding it through your everyday actions and decisions.

I would like to offer a thank you to Sabrina Meherally (Founder and Relational Design Strategist of **Pause + Effect**) who supported me in drafting the activity above! If you enjoyed this adapted activity, check out this incredible organization for more PD opportunities (pauseandeffect.ca).

Unveiling Your Teacher Disposition

By this point, your reflections, questions, and wonderings are unveiling more ideas and core values you hold about good science teaching. Teachers make decisions based on beliefs they hold about good science instruction and how students learn. When you take action on those beliefs, that is known as your *theory of action.* Just like beliefs, your theory of action may change over time as your beliefs change over time. You can see this in action by observing another colleague teach. During that observation, you will immediately notice that what works for your colleague, might not work for you. Similarly, what is acceptable in your class, might not be acceptable in your colleague's class. When you think about it that way, you begin to understand that your actions are based on underlying core teaching values that may differ from others, which is why they might have a different approach to a similar problem. I invite you to complete Exhibit 1.3 to better understand where you currently are with your instructional practices and NGSS.

Exhibit 1.3 Teaching Beliefs and Non-Negotiables

 Let's start with how you currently teach and your pedagogical decisions. *Think about the best science lesson that you have ever taught.* Answer the following questions to set the stage.

1. The best science lesson I ever taught was ...
2. How did that lesson start, what were the roles that you and students played, and how did the lesson end?
3. How did teaching that lesson make you feel?
4. If an observer saw your class, what would they see and hear during that lesson?
5. What were students capable of during that lesson?
6. Were there any challenges present during that lesson that called for refinement?
7. In order for this lesson to be successful, what are some actions that must take place or design considerations that must be included (we will call these **non-negotiables**)?

Pay close attention to #7 above. Pick one non-negotiable to focus on for this activity then reflect using the *"5 Whys"* method to help you uncover beliefs you hold about good science instruction. If you struggle to come up with clear rationale for why that is a practice you firming believe in, possibly reflect on that and how that makes you feel.

When reflecting on the best science lesson I have ever taught, one non-negotiable is...

1. *Why is that a non-negotiable?* **Because...**
2. *Why?* **Because...**
3. *Why?* **Because...**
4. *Why?* **Because...**
5. *Why?* **Because...**

If you completed this exercise for all your non-negotiables, what would you feel more affirmed about, consider changing, or question?

Limitations with the Scientific Method

Moving toward the new science framework, we must address the issues with *The Scientific Method*. Note that most teachers don't receive educational training explicitly on the Nature of Science through their teacher education programs or college experience, so they might be communicating about the scientific process as something that happens in a linear fashion. This is otherwise known as The Scientific Method. This method essentially lays out the process of science as an organized step-by-step instructional guide for students. The problem with this method is that it does not align with what scientists actually do or how they engage in scientific research. Science is about engaging in practices, *doing* science, and understanding the diverse range of approaches one can take to understand or solve a problem. One of the major issues with The Scientific Method is that it does not position students authentically as scientists because it fails to push them to make claims from evidence, modify claims/questions in light of new information, engage in discourse for sense-making, or acknowledge that people have different starting points in the process. The climate crisis and environmental issues that impact students' lives require complex and innovative ways of thinking. It's time to move beyond The Scientific Method, and teach students that science is something you do by engaging in a variety of practices through a variety of ways. Exhibit 1.4 supports you in reflecting on the rich Process and Nature of Science to begin thinking about updating your curriculum. Remember, the shift toward the NGSS requires a shift in thinking about how we teach and position students in the classroom.

Exhibit 1.4 A Better Understanding of the Process of Science

Many students are taught that science is about proving what we already know to be true. Contrary to that idea, science is actually about discovering new concepts, ideas, or solutions in creative ways. Students should be consistently asking questions and when presented with new evidence or information, ask even more questions. Think of The Scientific Method and The Process of Science as two goal posts. How can you shift your thinking in ways that reflect in your practices towards the more open-ended way of engaging in science? On the line below, notate where you were between these two concepts before you started this book, where you are currently, and where you want to end up.

The Scientific Method ----------------------The Process of Science

♦ Let's think back, when did you first learn about "The Scientific Method"? Do you remember if it was taught using a relevant example? What about whether (or not) it was a culturally relevant example? Whether it used a com-

plex problem of that time? Do you remember how engaging that experience was? Did it change the way that you thought about how the world functions? Did you question the process based on your own experience solving everyday problems?

♦ Are you confident in your knowledge of the Process of Science? What about how it shows up in the NGSS? How might it show up in your lesson design or unit planning? When do students typically engage in the Process of Science? The process emphasizes the need for students to drive sensemaking and inquiry starting with their own interests and ideas, how comfortable (or uncomfortable) might this be for both students and/or teachers? What scaffolds will be needed to build student capacity to drive the Process of Science? How might this process build students' scientific literacy skills to analyze issues that directly affect them?

Important considerations in the future when ready: There are many ways of knowing and being that are not emphasized in the NGSS that I would

like to lightly touch upon. One of which honors my own ancestors from Viet Nam, and how they cared for the land and nurtured the communities using Indigenous knowledge, sustainability practices, and wisdom. What are the ways we can develop students as complex critical thinkers capable of holding multiple truths in suspension with each other. How might that engage them in deeper ways? What supports might you need, to learn more about these culturally relevant approaches to deepen your connections with students and draw out their Funds of Knowledge?

To better visualize the Process of Science, UC Berkeley created an interactive tool (www.Bit.ly/CALTOOL2) that supports students in making their thinking visible as they explore science issues (see Figure 1.1). Looking at the visual, teachers should note how complex the process can be (as positive opportunities), how there are multiple entry points for students, and the variety of ways students can tap into their cultural wealth of knowledge to explore different topics. When thinking through ways to make your best lesson culturally relevant and responsive, you can use this tool to give your students more voice and choice with a central topic to investigate (see Appendix A on how to use the interactive tool). Exhibit 1.5 provides ideas on how to implement this tool with your students.

Exhibit 1.5 What Does This Look Like in the Classroom?

Teachers don't need to wait for the perfect moment to teach about the Nature and Process of Science. This could take place on a shortened day, or the first week of school as you are setting classroom norms/procedures/lab safety. Revisit the tool often to showcase the variety of methods students can employ to engage in science. Considering providing consistent opportunities for students to talk about their

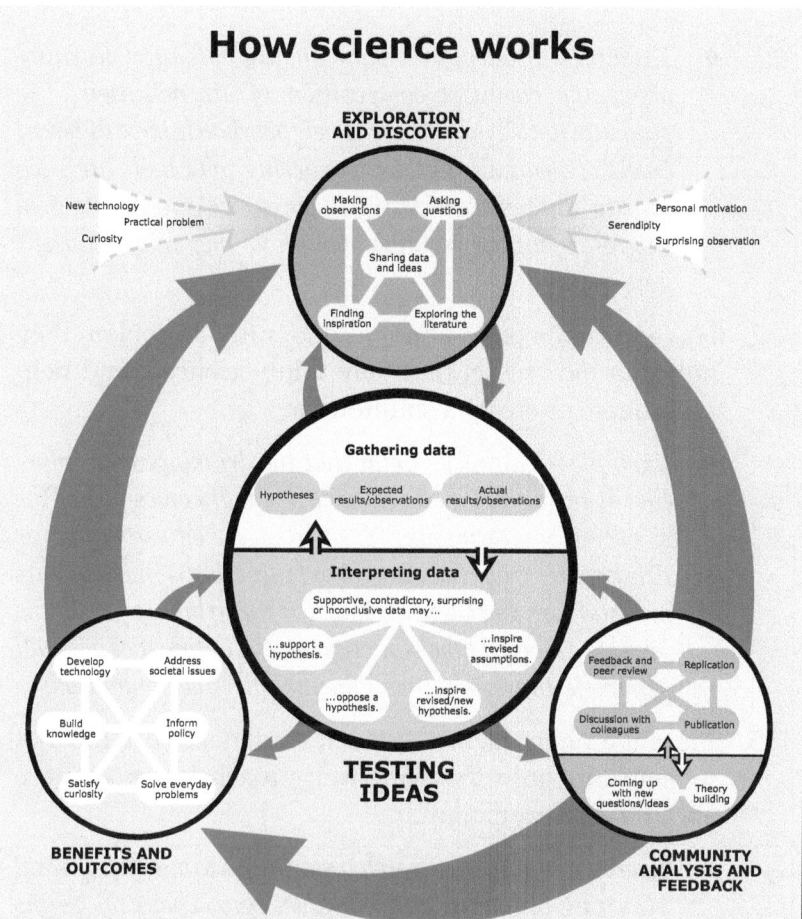

FIGURE 1.1 UC Berkeley's Process of Science Tool
Image Credit: Understanding Science by the University of California Museum of Paleontology.

maps and pathways to support sensemaking and ideation. This will help students to see each other as sources of information and capable contributors in your class and to the science field. See the following entry points to support students:

1. Have students think about and identify a problem they notice *in their neighborhood* that they think they could help investigate or create a solution for.

◆ *Example—Students view air quality data after learning about the chemical composition of air pollution. The students notice that different neighborhoods have different levels of clean air and are wondering why there are such drastic differences. Have students use the online tool in groups of three to think about how they might investigate this.*

2. Have students think about and identify a problem they notice *in their school* that they think science could help investigate or create a solution for.

◆ *Example—There's a rumor that the drinking water pipeline at the school is the same water that comes out of the science laboratory sinks. Students were advised to never drink water from the classroom faucet, but some students and staff are starting to wonder if the drinking fountains are safe as well. Have students use the online tool in small groups to think about how they might investigate this.*

3. Present a phenomenon and ask students to use the tool to map out how they might approach learning more about the phenomenon.

◆ *Example—Students watch a short 20 second clip of coral bleaching occurring in New Caledonia. The coral begins to change into neon blue, purple, and bright green colors rather than a pale white. Have students use the online tool in small groups to think about how they might investigate this phenomenon. Request that they chart out what questions they have about what they are seeing to help focus their investigation (allowing for different groups to pursue different interests).*

4. Have students keep track of how they engaged in science for a lab experiment they are conducting.

◆ *Example—Students just learned about the lack of access to clean water to nearly 1 million people on the planet and are exploring new water filtration methods to help address the problem. Students conduct a water filtration lab using nanofilters to remove three major contaminants (chemicals, urine, and bacteria). They compare results of water samples when carbon or zeolite are added to the nanofilters. Students are asked to bring a random sample of dirty water into class and have one chance to determine the best filtration method for their own sample. Have students use the online tool with their groups to think about how they might investigate this problem. Request that they chart out what questions they have about what they are thinking about to help focus their investigation for sense-making.*

Diving Deeper into the Process of Science

As we continue to think about ways to challenge the current model of science teaching and further align to NGSS, what will be needed is a paradigm shift between the old and new frameworks. The NGSS cannot be approached in the same way that teachers approached the older standards. The older framework was essentially a checklist of concepts that teachers needed to address prior to standardized exams. Rote memorization and constant recall of knowledge was valued and rewarded. Students were positioned as sponges who absorbed information that rarely connected larger ideas together. It's also important to note that the old framework ascribed to The Scientific Method, where specific steps were outlined with often predetermined answers at the end of the experience to gauge mastery. Designing culturally relevant and responsive lessons, unveiling students' current understandings, engaging in discourse, utilizing evidence to support claims, and much of the process of science was often negated with the old framework. If any of the new components took place, it was because the teacher invested personal time to learn more and had a strong desire to integrate them into the classroom reflecting their theory of action.

The NGSS is unlike any type of science instruction that we might be accustomed to or have personally experienced. This framework emphasizes engaging in science and engineering practices, cyclical and iterative ways of learning across grade levels and within content areas, stresses the need for climate education, and pushes for equity and justice-oriented ways of teaching. This is the moment for teachers to reimagine science education for what it could be. This directly connects back to teachers' beliefs about teaching and learning because you **have** the power to elevate certain principles and practices.

Paradigm shifts can be challenging and difficult because teachers must reflect on deeper levels first to make the it possible. Consider completing the following activity in Exhibit 1.6 to push your thinking and TOA based on what you've learned so far.

Exhibit 1.6 All Students, All Science?

What might be worth considering here is whether or not traditional approaches to schooling support every student to begin unveiling ideas of equity. Considering the following questions that may cause some discomfort (and take a moment to reflect on the reactions or questions that emerge in the margins):

♦ Who has benefited from the previous methods and opportunities of science instruction (old framework)?
♦ Who might not benefit or has been directly or indirectly harmed by previous methods of science instruction (old framework)?
♦ What does the research and history of education tell us about who is <u>consistently</u> disadvantaged and/or failing with previous science instruction?

Questions to think about the current science framework:

♦ Did you notice any issues with older science frameworks prior to the adoption of the NGSS? If so, what were they?

- In what ways does the NGSS challenge science instruction? In what ways does it provide student-centered instruction? In what ways might it still fall short?
- The new framework of science teaching calls attention to equity and real-world phenomena to anchor instruction. Why do you think those ideas are being addressed now? What has been your experience with teaching using these methods? What were your successes and challenges? What resources might you need?
- We know that science is something you do. In what ways do the older and new science frameworks align with this description of science? In what ways might it not?
- Have you looked up your state's NGSS test results yet (Bit.ly/CASTschool)? If so, are the results surprising, affirming, or upsetting? What do the results lack? What questions emerge for you when analyzing? Can understanding the old and the new frameworks (specifically focused on pedagogy and instruction) help you or your colleagues to align more to the NGSS?

 Check the box for the questions you are most interested in exploring and think about your current answers (They might reveal some of your teaching beliefs and rationale for instruction).

Getting Cozy with the Nature of Science

The NGSS highly promotes scientific literacy among students through essential topics such as climate change (NGSS Lead States, 2013). To support scientific literacy, the standards call for three-dimensional learning and push students to understand the Nature of Science (NOS). People need to be scientifically literate because they influence the implementation or formation of public science policies that may directly impact their lives. Individuals should be able to find, make sense of, or use information on topics of discussion to make well-informed decisions regarding science or technology related policies (Miller, 2016). Although 52% of Americans take interest in science or technology related issues, roughly 28% of the overall population is scientifically literate (Miller, 2016). Teachers are no doubt among this population, but accountability to NGSS requires that many need to quickly become scientifically literate to effectively support students. The first step is to look at major NOS themes called out in the framework.

The National Research Council (NRC) and the National Science Teachers Association (NSTA) both stress that a fundamental goal for science education is to produce scientifically literate people who understand the NOS (NRC, 2013; NSTA, 2016). There are **eight major principles** embedded within the NGSS to help students develop skills as informed critical thinkers and decision-makers. Table 1.1 highlights different NOS principles and how each relates to the NGSS Science and Engineering Practices or Crosscutting Concepts (view full table online by looking up NGSS Appendix H).

TABLE 1.1 Understandings about the Nature of Science (NRC, 2013)

NOS Principles	Middle School	High School
Scientific Investigations Use a Variety of Methods (SEPs)	• Science investigations use a variety of methods and tools to make measurements and observations. • Science investigations are guided by a set of values to ensure accuracy of measurements, observations, and objectivity of findings. • Science depends on evaluating proposed explanations.	• Science investigations use diverse methods and do not always use the same set of procedures to obtain data. • Scientific inquiry is characterized by a common set of values that include: logical thinking, precision, open-mindedness, objectivity, skepticism, replicability of results, and honest and ethical reporting of findings.

NOS Principles	Middle School	High School
Scientific Knowledge is Based on Empirical Evidence (SEPs)	• Science knowledge is based upon logical and conceptual connections between evidence and explanations. • Science disciplines share common rules of obtaining and evaluating empirical evidence.	• Science knowledge is based on empirical evidence. • Science includes the process of coordinating patterns of evidence with current theory. • Science arguments are strengthened by multiple lines of evidence supporting a single explanation.
Scientific Knowledge is Open to Revision in Light of New Evidence (SEPs)	• Scientific explanations are subject to revision and improvement in light of new evidence. • Science findings are frequently revised and/or reinterpreted based on new evidence.	• Scientific explanations can be probabilistic. • Most scientific knowledge is quite durable but is, in principle, subject to change based on new evidence and/or reinterpretation of existing evidence. • Scientific argumentation is a mode of logical discourse used to clarify the strength of relationships between ideas and evidence that may result in revision of an explanation.
Science Models, Laws, Mechanisms, and Theories Explain Natural Phenomena (SEPs)	• Theories are explanations for observable phenomena. • Science theories are based on a body of evidence developed over time. • A hypothesis is used by scientists as an idea that may contribute important new knowledge for the evaluation of a scientific theory.	• A scientific theory is a substantiated explanation of some aspect of the natural world, based on a body of facts that has been repeatedly confirmed through observation and experiment, and the science community validates each theory before it is accepted. If new evidence is discovered that the theory does not accommodate, the theory is generally modified in light of this new evidence. • Models, mechanisms, and explanations collectively serve as tools in the development of a scientific theory. • Laws are statements or descriptions of the relationships among observable phenomena.

(continued)

TABLE 1.1 *(continued)*

NOS Principles	Middle School	High School
Science is a Way of Knowing (CCCs)	• Science is both a body of knowledge and the processes and practices used to add to that body of knowledge. • Science knowledge is cumulative and many people, from many generations and nations, have contributed to science knowledge. • Science is a way of knowing used by many people, not just scientists.	• Science is both a body of knowledge that represents a current understanding of natural systems and the processes used to refine, elaborate, revise, and extend this knowledge. • Science is a unique way of knowing and there are other ways of knowing. • Science knowledge has a history that includes the refinement of, and changes to, theories, ideas, and beliefs over time.
Scientific Knowledge Assumes an Order and Consistency in Natural Systems (CCCs)	• Science assumes that objects and events in natural systems occur in consistent patterns that are understandable through measurement and observation. • Science carefully considers and evaluates anomalies in data and evidence.	• Scientific knowledge is based on the assumption that natural laws operate today as they did in the past and they will continue to do so in the future. • Science assumes the universe is a vast single system in which basic laws are consistent.
Science is a Human Endeavor (CCCs)	• Men and women from different social, cultural, and ethnic backgrounds work as scientists and engineers. Scientists and engineers rely on human qualities such as persistence, precision, reasoning, logic, imagination and creativity. • Scientists and engineers are guided by habits of mind such as intellectual honesty, tolerance of ambiguity, skepticism and openness to new ideas.	• Scientific knowledge is a result of human endeavor, imagination, and creativity. • Individuals and teams from many nations and cultures have contributed to science and to advances in engineering. • Scientists' backgrounds, theoretical commitments, and fields of endeavor influence the nature of their findings.

NOS Principles	Middle School	High School
Science Addresses Questions About the Natural and Material World (CCCs)	• Scientific knowledge is constrained by human capacity, technology, and materials. • Science knowledge can describe consequences of actions but is not responsible for society's decisions.	• Science and technology may raise ethical issues for which science, by itself, does not provide answers and solutions. • Science knowledge indicates what can happen in natural systems—not what should happen. The latter involves ethics, values, and human decisions about the use of knowledge. • Many decisions are not made using science alone, but rely on social and cultural contexts to resolve issues.

Content Credit: NGSS Lead States (2013) (https://www.nextgenscience.org/).

Note: NGSS Appendix H is a registered trademark of WestEd. Neither WestEd nor the lead states and partners that developed the Next Generation Science Standards were involved in the production of this product, and do not endorse it.

Equity and Antiracist Science Teaching

Although science can effectively tell us what and how to teach regarding climate change, we cannot ignore social science data revealing why changes are still not taking place. Part of unveiling our underlying belief system means that we need to take time to sit uncomfortably with how those beliefs influence our actions and decisions with possible bias. To support your thinking, I'm listing a few of the harmful beliefs shared by some of the most well-intentioned educators I've worked with:

◆ *"English is the superior language."*
◆ *"The educational system works for everyone as long as they try hard enough."*
◆ *"Historically marginalized students don't have the rigor needed to pursue or succeed in STEM careers."*

Carefully consider how these deficit core beliefs might influence a teacher's everyday pedagogical decisions (regardless of whether or not they are aware of it). "Good intentions do not negate impact," and we need time to unveil our core beliefs that shape our everyday teaching. Remember that the educational system is made and reaffirmed by people. If we want to disrupt that system in any way, we have to start by looking inwards to see if we are perpetuating the cycle or actively working to dismantle it. If we start with ourselves and our own spheres of influence, we can create a rippling effect of change.

Ibram X. Kendi (2019) asserts that when it comes to racist or antiracist work, we are all actively engaging in one or the other through our daily decisions. Recall that teachers hold a great deal of power in the classroom when they are able to assign grades, choose which voices to elevate, design curricula, create community, and much more. As we reimagine what science education can be, think about ways that your daily lessons affirm your students' viewpoints, culture, or identity. When you are ready to challenge yourself to critically reflect and work toward equity and antiracist science teaching, consider engaging with the activity in Exhibit 1.7.

Exhibit 1.7 Teaching Science as Antiracists

As you read through and think critically about your own practices and beliefs, indicate which questions you find yourself drawn to and want to know more about. Consider the following umbrella questions with supportive guiding questions to engage in antiracist science teaching:

◆ Where do my students see themselves in this lesson?

  ◆ Specifically, how are they represented or positioned during today's lesson?
  ◆ At what point(s) will they co-construct knowledge with me to make sense of what they are learning?
  ◆ Who is driving this lesson (me, my students, both)?
  ◆ Is the lesson culturally relevant or responsive in any way?

◆ How does this lesson affirm students' cultural identity to empower them by positioning them as capable contributors?

   ◆ When I teach this lesson, are students positioned as empty buckets that need to be filled with content knowledge, or as individuals with cultural wealth and knowledge that can contribute in different ways?
   ◆ Do I know my students' backgrounds or family histories? Does it make a difference in how they decide to engage/disengage in class?
   ◆ How might storytelling be empowering or affirming for students in my class?
   ◆ Do my students see themselves represented in the classroom space authentically or performatively?
   ◆ How important is genuine relationship building in my class? Why?

◆ Do I recognize ways in which I have assimilated to the dominant culture, and how that may potentially be shaping my pedagogical practices and classroom decisions?

   ◆ First, am I aware of teaching practices that reinforce the dominant culture? Try thinking of a scientist or engineer. What do they look like? What do they sound like? Do all scientists or engineers look like the person I'm envisioning?
   ◆ Are there benefits to my curriculum being more culturally relevant and responsive for my students?
   ◆ Is there value in affirming students' cultural identities, their unique ways of thinking, or their different ways of doing science?
   ◆ When I showcase scientists or engineers, are they diverse and from different cultures around the world?
   ◆ Do I hold scientists from my country in higher regard than scientists from other countries? Does this shape my viewpoints or reflect beliefs I hold?

Kendi further defines antiracists as individuals who believe that racial groups are equals that **do not need developing** (2019). You might be wondering how that translates in a classroom setting. It always begins with teachers who have **not** taken the time to explore their underlying core values and beliefs. In this example, you have teachers unintentionally (or intentionally) pressing upon students the need to assimilate for better life opportunities (i.e. succeeding in a job, going to college, conforming to specific societal rules, etc.). They decide to teach underrepresented minority (or the global majority) students how to "code-switch" without any historical context. Code-switching is when the teacher stresses the need to speak or behave in specific ways in the classroom to blend in or be accepted. On the surface they mean well when they ask students to correct the way they express their ideas to "sound like scientists," but how might that shape students cultural or STEM identity if they heard that multiple times per day by various teachers? Why is *what* they are saying discounted because of the *way* they're saying it (even when the ideas or questions posed are valid)? When they're repeatedly told to change part of themselves, what is the underlying message about who they are, how teachers might see them, or whether or not they belong in the STEM fields?

STEM fields *need* more diversity for different perspectives and ideas to advance for the betterment of our planet. Antiracist teachers are those who believe that racial groups are capable co-contributors with a wealth of knowledge. Code-switching does not affirm students' identity or validate who they are—instead it is about developing them into who we *believe* they need to be to obtain success in our society (now or later in life). The fact is, many scientists and engineers do not use formal ways of articulating ideas, may or may not speak or even write in English, are diverse from around the world, and communicate scientific studies and inventions in various celebrated languages and ways. We need to think about how these subtle lessons (that have nothing to do with science and everything to do with culture) might be serving as barriers to historically marginalized students and STEM.

How empowered would students feel if they knew how cultures around the world have contributed and continue to contribute to STEM fields? For example, look at the awe-inspiring science and engineering designs that complement nature in Micronesia (an under-appreciated third-world country). Back in the year 1200, The Saudeleur Dynasty manipulated inland waterways to transport large basalt stones (some weighing 100,000 pounds) across 25 miles to create the engineering wonder that is Nan Madol (a lost city where the wealthiest and most revered once inhabited as they controlled oceanic trading operations). Advanced science and engineering took place here before the words "science" and "engineering" even came into existence. How we think and feel about our students and their capacities will impact how and what we teach. Are we able to position them as capable contributors while building their capacity, while celebrating their "whole selves" to achieve what others might believe is impossible?

If we want to reimagine what science teaching can be to support every student, Zaretta Hammond argues that, "We need to be brave enough to interrogate our practice for the sake of our kids (2015)." Just as building relationships with students takes time through small daily interactions, creating an inclusive classroom space that serves as both "windows" into other cultures as well as "mirrors" that reflect students requires intentionality (Sims Bishop, 1990). Being an antiracist is not a destination, but rather a *process* you engage with as you make classroom decisions ranging from how to deliver a lesson, its intentional design, and creating the space in which it will all take place.

Putting the Pieces Together

If the previous section made you feel any discomfort because it challenges some pre-existing ideas or beliefs you hold, take a moment to process and discuss your thoughts with someone you trust. Then take time to learn more about the history and purpose of the education system for questions you still hold. When you're ready, let's help you continue to connect the dots. We will now parallel your learning experience to that of your students to shed light on similarities and differences when learning new science ideas.

Educational researchers emphasize three main elements needed to support students learning science (National Academies of Science, Engineering, and Medicine, 2018; NRC, 2005). The first element is that teachers need to address students' current understandings about the content (some might refer to this as prior knowledge). If you want to teach students new content, you have to first unveil what they currently know and why/how they come to know that information. If there are initial misunderstandings, consistent opportunities for discourse will let you gauge their depth of knowledge and potential gaps to use as starting points. Remember that if a student has firmly held on to certain misconceptions for long periods of time, it will take more than one class period for them to believe they are wrong just because you say they are. Exhibit 1.8 will provide you with various teacher talk moves to support this element.

Exhibit 1.8 Intentional Teacher Talk Moves

Teachers are more likely to support students' ability to connect their current understanding with new concepts by being intentional in their questioning, listening, and affirmation of students' thoughts. This happens effectively through consistent and multiple opportunities for discourse to allow for co-construction of knowledge with students.

Deliberate Questioning

◆ *Encourage students to say more.* This move allows for teachers to gauge how many students feel or think similarly, and allows them to reflect more deeply about what they are thinking.

◆ *Revoice responses.* By reiterating what the student said, it could bring clarity to the student who hears it back so they can agree or correct their own ideas.

♦ *Ask for reasoning using evidence.* Go beyond a superficial level of conversation to deeper discourse for sense-making by exploring why students think or feel that way about their adopted beliefs.

Listening With Intent

♦ *Powerful paraphrase.* Ask students to restate what their peers are sharing to engage students in conversation and allow for students to affirm each other's ideas.
♦ *Challenge or offer counterexamples.* Are you or students able to push each other by providing counterexamples? Consider also trying this with a student who provides the right answer to unveil their thinking and rationale.
♦ *Wait time.* Provide wait time between questions and responses so students can process information at their own pace. Consider tracking how often you reward the fast thinkers in your class and how to more equitably tend to every learner.

Affirming Students

♦ *Acknowledge.* Understand that your students have these current ideas that stemmed from something very real to them. Start where they are and through questioning, bridge the gap so they can get to where they need to be. Is there something in their response that allows for you to empathize with *why* they adopted that idea?
♦ *Set up Affirmation Opportunities.* Ask a student to share and find students who will affirm the initial student's ideas. This will allow you to see who else shares the same current understanding and why. When you move to asking who disagrees and why, it would be helpful to gently point out that it's okay to be wrong (engaging in the Nature of Science) because perhaps they didn't have the information or data currently presented and more importantly, they are not alone in how they thought about that old idea and it's understandable why they did.

The second element emphasizes the need for students to understand the Nature and Process of Science. Moving beyond The Scientific Method, this element calls for students to be positioned as scientists and engineers by engaging with different practices to do science. If we are relying on students to use science as a tool to solve real-world complex problems, they need to be scientifically literate and have skills to apply their knowledge. Recall that positioning students as scientists and engineers will greatly depend on your teaching disposition. Take time to reflect on your values and beliefs about how students learn in relation to equity and antiracist teaching practices. Participate in Exhibit 1.9 to tend to equity in your class through challenging reflection.

Exhibit 1.9 Tending to Equity in Science

The following lesson was created in partnership with Ashley Herrmann formerly at Our Climate Our Future to highlight energy justice. The lesson uses the Department of Energy L.E.A.D. tool to support student engagement and exploration around a complex issue with the goal of supporting sense-making, decision-making, and to activate their agency starting at the community-level. Access the full lesson with video support at bit.ly/3wCUdcJ (case-sensitive).

Overview: Through a short demo, students will first investigate and ask questions about where energy comes from and how it powers things by watching a lightbulb illuminate near a plasma ball without being plugged in to activate their curiosity! As a facilitator, the teacher will support students in determining an anchoring *class* question. All students' questions will generally relate to one of the following: "What is energy?" or "Where does it come from?" or "How does it power something?" (for tips see "Approaches to Support Students Developing the Driving Questions" section in Chapter 5). To support the class in determining *one* anchoring class question from all the questions they presented, considering posing the following as guidance:

◆ Which question, *that if I answered it in its entirety*, would explain everything there is to know about this phenomenon?

   ◆ *Tip:* The phenomenon is the lightbulb turning on without touching the plasma ball.
   ◆ *Tip:* Guide the students to selecting a question general enough that it would allow for them to explore multiple perspectives and related content to fully understand what is going on with the lightbulb.
   ◆ *Tip:* Take one of the students' questions that is specific with a yes/no response and ask it out loud to the class. Then provide the answer of either "yes" or "no," and ask if they feel they have understood the entire phenomenon now that you gave them the answer. *Example—Student question: "Does the energy come from the plug connected to the plasma ball that leads to the wall?" Teacher Answer: "Yes." "Now that you know the answer, does that explain everything that you need to know about how the lightbulb is on without touching the plasma ball?"*
   ◆ *Tip:* Over the years I learned that there are many ways to support student reflection and metacognition. Once they determine the anchoring question, I let them know that it will be the short answer essay question on the unit exam. This empowers them to learn as much as possible and sense-make throughout the weeks to be able to explain the phenomenon through written, oral, and artistic forms. Their *personal* question about the phenomenon (it might not have anything to do with the anchoring class question because it's extremely specific), becomes the second short answer question to honor what they want to learn more about and encourage exploration on their own outside of what we might cover.

After some fun with this section, introduce students to the LEAD interactive tool by allowing them to explore all the features first without guidance and sharing those features with a peer. This is where you would teach about what energy is, sources of current energy powering things (content specific information about harmful sources of energy and disproportionate levels of harm discussed here), and which sources are primarily used across specific communities (as shown on the LEAD tool).

The lesson ends with a deep discussion with students on thinking about energy burden scores across various communities (including their own the communities neighboring them) to provide time to discuss and tend to social and emotional needs that will likely emerge as a result of seeing data and making sense of their disproportionate lived realities. This also provides great extension opportunities to support students in respectful argumentation in science and to activate their agency by sharing their knowledge and thinking about ways the information can help with decision-making starting at home with their families and community members. Access the full free lesson with video support to help students with those "lightbulb" moments today in ways that engage them with the Nature of Science and NGSS practices.

The third element highlights the need for students to participate in metacognition (thinking about their own thinking for sense-making). Allowing students to reflect and think about their thinking fosters cognitive growth. Much of the NGSS requires students to provide reasoning or justification for a claim. Allowing students to question themselves, their decisions, or their peers provides opportunities for them to be critical consumers of information. Science relies heavily on credible sources of evidence that allow individuals to accept, reject, or modify claims in light of new information. When students see their teachers or peers modeling the NOS and Process of Science, they begin to understand how credible evidence carries more weight than any number of expert opinions. Consider providing opportunities for students to identify their own learning dispositions when you start class. Can you tend to their social and emotional well-being, so they have moments to recognize what motivations or challenges are impacting their learning experience today?

Without analyzing your underlying core beliefs and teaching values, you might inadvertently approach the NGSS in ways similar to the old science framework. To ensure that you are successful in taking on climate change for NGSS, consider what you're willing to change, why you would invest time to change it, and for what purpose that change was made (revisit the "5 Whys" in Exhibit 1.3). Just as the NGSS is not a series of checkboxes that may or may not fit into your current curriculum, teaching about climate change to empower students requires more than just temporary strategies or tools. For both, there needs to be a paradigm shift. This chapter is meant to provide you with an overview of what is needed to enhance your science practices to align to NGSS, provide a clear rationale for the paradigm shift, and offer ideas to consider while you reflect on your theories of action.

 *Connect with the network of teachers to share your thoughts on education for climate action by leveraging the NGSS at www.EmpoweredScienceTeachers.com (Book Resources → Chapter 1 → Discussion Board)*

Collective Voices for Climate Change Education

Shraya Sharma, Manager of Content at Empatico

*Our students will be responsible for solving some of the world's most complex issues, such as climate change, environmental sustainability, and inequitable access to natural resources. In order to tackle these challenges, we must equip our students with the mindset, skills, and will to take compassionate action locally and globally. At Empatico and the UC Irvine Science Project, we are determined to **foster a generation of empathetic problem solvers** who show curiosity about others' perspectives and ideas, care about others' feelings and experiences, and are able to communicate and collaborate respectfully. They not only consider how these challenges will impact them and their local communities, but they also understand and care about how these challenges will impact people living in communities around the world. Together, we can inspire the next generation of empathetic leaders who will shape a brighter and more sustainable future for everyone.*

Ashley Herrmann, Director of Education formerly at Our Climate Our Future

Young people have the right to know about the climate emergency and must be supported to take action to stop it. I had the privilege of working with Dr. Kelley T. Lê to create six lesson plans that are offered for free on https://ourclimateourfuture.org/resources. Each incorporates UDL and addresses NGSS. Several were selected by CLEAN (Climate Literacy and Energy Awareness Network) to feature as high-quality resources. Kelley T. Lê's book is an incredible companion to anyone who wants to understand and teach others about the complexities of climate change.

Additional Teacher Resources

Access "How Students Learn (2005)"—
Bit.ly/HSLSCIENCE
Access Our Climate Our Future teaching resources—
Bit.ly/49WXUrX
Diversify Our Narrative Educator Resources—
Bit.ly/DONTOOLKIT
Explore The Great Empatico Expedition for K-8 to tend to
SEL—Bit.ly/4c1FDM4
Get more teaching tools with UC Berkeley's Resources—
Bit.ly/CALTOOL2
Launch a K-12 eco-lesson with a grant through
Grades of Green—www.Bit.ly/GRADESOFGREEN
Learn more about the Nature of Science in the NGSS—
Bit.ly/NGSSNOS
Learn more about the Process of Science—
Bit.ly/ProcessOfScience
See your school's CAST results here—
Bit.ly/CASTschool

References

Anderson, A., Druger, M., James, C., & Katz, P. (1998). An NSTA position statement: The national science education standards: A vision for the improvement of science teaching and learning. *Science and Children, 35*(8), 32.

Duggan Haas, D. (2020). *Science in the virtual pub.* Paleontological Research Institution. https://www.priweb.org/event/science-in-the-virtual-pub

Kendi, I. X. (2019). *How to be an antiracist.* First ed. New York: One World.

Hammond, Z. (2015). *Culturally responsive teaching and the brain: Promoting authentic engagement and rigor among culturally and linguistically diverse students.* Thousand Oaks, CA: Corwin.

Miller, J. D. (2016). Civic scientific literacy in the United States in 2016. *International Center for the Advancement of Scientific Literacy: Ann Arbor, MI, USA.*

National Academies of Sciences, Engineering, and Medicine. (2018). *How people learn II: Learners, contexts, and cultures.* Washington, DC: The National Academies Press. https://doi.org/10.17226/24783.

National Research Council. (2005). *How students learn: Science in the classroom.* Washington, DC: The National Academies Press. https://doi.org/10.17226/11102.

National Research Council. (2013). *Next generation science standards: For states, by states.* Washington, DC: The National Academies Press. https://doi.org/10.17226/18290.

NGSS Lead States. (2013). *Next generation science standards: For states, by states.* Washington, DC: The National Academies Press.

Sims Bishop, R. (1990). Mirrors, windows, and sliding glass doors. *Perspectives: Choosing and Using Books for the Classroom*, 6(3), ix–xi.

2

Leveraging the NGSS for Climate Education

Read this when:

♦ *You're ready to engage in more NGSS-aligned instruction.*
♦ *You're considering the role of climate change in the NGSS and wondering about how the two complement each other.*
♦ *You're ready to put more into practice.*

Critical Perspectives for Complex Thinking

Living in Los Angeles, I'm accustomed to spending a great deal of time on the freeways. To pass the time while on the road, I'm always looking around to see if I can spot wildlife or signs of a fire. On more than one occasion, I've driven on the 405 freeway seeing fire burn the dried up areas that frame the roads. I was taught by both society and schools that wildfires are bad. So much so that when I smell or see the ash falling from the larger fires (that now come during anticipated "seasons" because they're so normalized), I think of danger, destruction, death, negative human health impacts for those who have respiratory illnesses, and more. I learned that fire was scary, bad, and uncontrollable.

DOI: 10.4324/9781003478584-4

That narrative became more complex for me when I attended Lyla June's workshop as a RegenIntel Fellow, and learned about the *regenerative medicine of fire*. Lyla taught us about how,

> Fire is one of the most important tools for regenerating soil, maintaining meadows, spacing trees, reducing thicket of shrub and mesic forests, and keeping the land clean and pretty and cycling nutrients.... In the wake of fires, the area that was burned would have nutrient dense grasslands that would attract animals like buffalo, deer, and such.
>
> (2024)[1]

She taught us about a large-scale Indigenous regenerative system in the United States, that relied on fires to effectively manage lands called the Native American Grassland Pyro-management System. This system supported life and regeneration for thousands of years before colonization prohibited these practices. With the intentional burning of dryer lands every fall season to maximize productivity, came the production of nutrients such as phosphorus, nitrogen, soil, and the original biochar that people could use to stimulate microbial activity to filter water and more. Lyla goes on to share,

> A lot of us are scared of fire, but we don't understand that from an Indigenous lens, fire is medicine. Fire actually opens up meadows. Fire feeds the soil. Fire renews the land, and the land has come to expect our fire because without us it will take fire through catastrophic fire. Our fires actually prevent all those scary fires we see. Those fires are actually caused by the prohibition of Indigenous fires as early as when the Spanish missionaries were there before it was part of America.
>
> – Lyla June

Everything I know and feel about fire is what I learned in and beyond schools and then also witnessed. I realize now, however, that I was given an incomplete picture, which shapes and influences my thinking and my decision-making. It's never too late to learn new things right?

This is a prime example of how to teach students complex critical thinking skills to hold multiple ideas in suspension. Given that there are many uncontrollable external factors and challenges that also play a role in students' educational trajectory, there is no denying the powerful difference that every teacher (both formal and informal) makes. What students learn and develop in your science class is also determined by what you are ready to take on or what you decide to push off. Based on your personal beliefs, you have the power to determine if students are capable of driving the lesson, will engage in meaningful discourse, share their cultural wealth, engage in STEM, learn more about Indigenous knowledge, and more. This chapter will continue to deconstruct the NGSS to leverage the framework using climate science as the vehicle to support your successful classroom implementation of the material.

A Second Look at the NGSS for Climate Science

When thinking about the shift to NGSS, we need to accurately identify the problem in order to come up with the right solution. When you listen carefully to how people frame the issues regarding NGSS, it reveals a great deal about their values and beliefs that influence their actions. Without a paradigm shift, teachers might inadvertently approach the NGSS as disconnected components they force into current lessons. Although it is the easiest approach, we know this type of approach will not get us the results we are hoping for. More importantly, this type of approach does not take into consideration the pedagogical shifts needed to implement NGSS curricula not generally included in lesson plans. Anyone can pull "NGSS" resources off the internet, but how they enact the lesson reveals another level of alignment needed for the framework. It can be an exciting professional challenge to revamp your curriculum, especially if you're thinking of using climate change as the driving force to engage and empower students. When climate literacy is culturally and locally relevant to students, students have authentic opportunities to develop confident scientific identities grounded in solution-oriented thinking with skills to apply to the problems of today. Participate in Exhibit 2.1 to see if any of your beliefs about NGSS have shifted.

Exhibit 2.1 Why the Shift to NGSS?

Forty-four states have either adopted the NGSS or have education standards heavily influenced by the new framework since 2011, moving away from the former national science framework that did not include climate science or engineering practices. Review the list below and select the *top three reasons* for why you think that shift might have happened.

- ☐ The standards change every decade or so, it was just a matter of time.
- ☐ There were too many content requirements in the old framework and students needed more depth in their learning.
- ☐ Because *doing* science is essential for students, which is emphasized in the NGSS.
- ☐ Nationally, students were not as prepared as they could be to persist in STEM fields.
- ☐ We have more research revealing new information on how students learn science today (e.g. including culturally relevant and responsive pedagogy, considering the role of culture and learning, tending to students' social and emotional well-being, centering on equity, etc.).
- ☐ There were resource inequities between schools accountable to the same standards.
- ☐ There is a need for more coherence between and across science classes.
- ☐ Students needed more authentic opportunities to engage as scientists and engineers.

☐ Students needed to start developing scientific literacy skills.

☐ Students needed more opportunities for sense-making rather than memorization of facts.

☐ Other: _____

☐ Other: _____

Depending on the reasons you identified, that reveals the lens that you might be approaching teaching with. For example, if I *only* checked the first box I might be thinking that the NGSS is just another state requirement. I might not also value the components of the NGSS because I view it as another requirement to get through (a change just to change something rather than having a meaningful purpose). I also might be thinking that "This too shall pass," so if I wait it out long enough new standards will replace this one so there's no point in revamping my curriculum. I also might not see flaws in the former framework that teams of leaders decided to adopt because the majority of my students did well in "mastering" the content. You can see how these beliefs will shape the actions that are taken by the teacher (a.k.a. their theory of action) in shifting toward the NGSS.

 Think about your values and beliefs about teaching and learning. What actions might you take depending on the beliefs you have about shifting to the NGSS?

Think about your science colleagues. What actions do you think they are taking regarding the NGSS, which might reveal some of their beliefs about the new framework?  Thinking about this will help you to better understand how others might operate and lead to better collaboration efforts in the future if you can find middle ground.

This book focuses on the following three reasons for the shift to NGSS that includes (but are not limited to) the following:

1. Treating science as a creative and innovative process to solve everyday problems, rather than only rediscovering what we already know to be true.
2. Using climate change as the vehicle to learn about the Nature of Science to further develop students' scientific literacy skills.
3. Positioning all students as knowledgeable and capable doers of science that can contribute to STEM fields.

Approaches provided by this book to address the above include education on the Process and Nature of Science to shift the way we think about science teaching through NGSS for climate change (Chapters 1 and 2), beginning support on integrating climate science into your curriculum for scientific literacy skills (Chapters 3 and 4), and providing advanced resources to empower your students to use science to address issues in their communities in which they are well positioned to help lead (Chapters 5 and 6). To also have a better understanding between the old and new science frameworks, review Table 2.1 for an overall comparison.

A Fresh Take on the NGSS

The new science framework calls for teachers to integrate science content and to support the generation of solutions to combat societal problems (the perfect catalyst needed for education leading to climate action). There are three major components of the NGSS that include: the science and engineering practices (SEPs), disciplinary core ideas (DCIs), and crosscutting concepts (CCCs). The SEPs refer to what scientists do as they investigate phenomena around them and how engineers design or create systems as a response. The DCIs are key content ideas that build off one another through and across grade bands. Lastly, the CCCs are the connections between major science domains (such as Earth and Space Science, Physical Science, Life Science,

TABLE 2.1 Comparing the Frameworks

Previous Science Framework	New Science Framework (NGSS)
1. Held students accountable for a wide range of content.	1. Holds students accountable to performance expectations that calls for three-dimensional learning.
2. Rewards students skilled at rote memorization.	2. Positions students as capable critical thinkers, scientists, engineers, and agents of change.
3. Thinks of student learners as sponges that mainly absorb content information.	3. Thinks of students as knowledgeable contributors and values their cultural backgrounds and communities.
4. Excluded engineering.	4. Integrates engineering components.
5. Excluded climate science.	5. Integrates climate science.
6. Stresses The Scientific Method.	6. Introduces the true process of science.
7. Often assessed through multiple-choice emphasizing content recall.	7. Emphasizes scientific literacy through learning about the Nature of Science.
8. Students are not accountable for applying content knowledge to the real world.	8. Assesses students' critical thinking skills to apply content rather.
9. There's no emphasis on cross-curricular or interdisciplinary teaching.	9. Recognizes the complexity of science and calls for an integration of mathematics and English language arts.
10. There is little accountability and unclear expectations for what students will do with the content.	10. Aims to prepare all students for college, career, and citizenship.
11. Does not consider research on how students learn new science content.	11. Takes into account how students learn through cyclical and iterative processes (example: sense-making through phenomenon-based instruction).
12. Lacks culturally relevant approaches to teaching science.	12. Views culturally relevant pedagogy as essential to engaging and empowering students.
13. Emphasis is on knowing about science.	13. Emphasis is on doing science.

and Engineering). When seamlessly intertwined, the three components provide students with three-dimensional learning (Refer to Table 2.2). This approach supports building students' capacity as well-informed decision-makers on STEM policy by increasing their scientific literacy skills.

Nearly one-third of the NGSS relates directly or indirectly to climate change content, and there are many opportunities to

TABLE 2.2 NGSS Three-Dimensional Learning

Science & Engineering Practices	Disciplinary Core Ideas	Cross Cutting Concepts
• Ask questions and define problems • Plan and carry out investigations • Use mathematics and computational thinking • Engage in an argument from evidence • Develop and use models • Analyze and interpret data • Construct explanations and design solutions • Obtain, evaluate and communicate information.	• Life Sciences • Physical Sciences • Earth & Space Sciences • Engineering, Technology, and Application of Science	• Patterns • Scale, proportion, and quantity • Energy and matter • Cause and effect • Systems and system models • Structure and function • Stability and change

Content Credit: NGSS Lead States (2013) (https://www.nextgenscience.org/).

Note: The NGSS is a registered trademark of WestEd. Neither WestEd nor the lead states and partners that developed the Next Generation Science Standards were involved in the production of this product, and do not endorse it.

incorporate climate science coherently within and across grade bands. Since the majority of teachers didn't receive formal training on teaching climate science for NGSS, it's important to highlight climate science as it relates to the new framework. Although not explicit in elementary standards, primary teachers do teach about the cycling of matter and Earth's spheres as a basis of climate science among other concepts. For secondary standards, climate science is written through different Performance Expectations (PE) and DCI (e.g. including Weather and Climate, Human Impact, Human Sustainability, Earth's Systems, Earth & Human Activity, etc.), energy is a CCC woven throughout all courses, and pushing students to develop solutions for climate-related issues they will face requires the SEPs and Engineering, Technology, and Application of Science standards (ETS). Please refer to *Additional Teacher Resources* at the end of the chapter to access several resources that outline climate science in the NGSS.

Along with understanding where climate science is embedded in the NGSS, it's also important to note that young people across the country want support in learning about and leading change to fight the climate crisis. According to the US National Action for Climate Empowerment Strategic Planning Framework, public empowerment is necessary to meet the ongoing challenges of climate change (Bowman & Morrison, 2020).

> *The solutions to the negative effects of climate change are also the paths to a safer, healthier, cleaner and more prosperous future for all. However, for such a future to become reality, citizens of all countries, at all levels of government, society and enterprise, need to understand and be involved.*
>
> (Paas & Goodman, 2016)

Pressing issues such as the climate crisis are predicted to lead in education as environmental justice issues become the center of attention this century. With this in mind, climate change serves as the ideal topic because it directly impacts people and will require innovative, diverse, and creative ways of thinking to mitigate.

The Role of Climate Literacy & NGSS

Although climate science is complex, interdisciplinary, and multi-faceted, there are major scientific components that every science teacher should address with students that have been identified by NGSS and researchers. The NGSS Earth and Space Science PEs encourage students to…

- ◆ understand system interactions that influence weather and climate,
- ◆ analyze and interpret geoscience data that drives climate change,
- ◆ unpack the significant interdependencies between humans and Earth's systems by looking at the impacts of natural hazards,

- ◆ critically examine our dependency on natural resources,
- ◆ and recognize the impact human activities have on the environment.

Students demonstrating content mastery should be able to develop and use models, analyze and interpret data, apply mathematics and computational thinking, construct explanations, design solutions, and engage in argumentation using evidence. The development of this section in the NGSS was strongly influenced by several literacy principles including the NOAA Climate Literacy Principles (2009), which outlines major learning objectives for students.

Similar to the NGSS, *The Teacher-Friendly Guide to Climate Change* (TFGCC) identified five major concepts that all students should understand about climate change. The authors interviewed expert climate scientists, social scientists, science educators, and climate journalists who agreed on big ideas along with two overarching questions educators should consider when teaching climate change (Zabel, Duggan-Haas, & Ross, 2017). They also acknowledge the importance of starting with the Climate Literacy Principles to focus on what all students should understand as a basis for climate science. The **first** big idea is to acknowledge that climate change is a real and very serious problem that our society faces now and in the coming centuries. **Second**, climate change is caused by anthropogenic factors, especially when it comes to energy use. Anthropogenic is a term used to describe the impact on Earth as caused or influenced by human activities. **Third**, it is important to understand that humans can take actions to mitigate climate change and its impacts. Fourth, there is a need for mathematical thinking to understand time, scale, models, and maps in depth related to climate change. Lastly, experts argue that understanding that Earth is a system of complex systems is the most important concept because many subjects are connected in explaining the climate crisis (special note that the four big ideas mentioned previously rely on this one). The two overarching questions draw upon the Nature of Science (NOS) to encourage students to examine how scientists know what they know to understand the scientific process, and how that information informs decision-making. Table 2.3 outlines the Climate Literacy Principles, climate change content in the NGSS, and the recommendations put forth by the TFGCC to show points of intersection.

TABLE 2.3 Connections for Climate Change Curriculum Design

Climate Literacy Principles (NOAA, 2024)	Next Generation Science Standards (NGSS, 2020)	Teacher-Friendly Guide to Climate Change (PRI, 2017)
Essential Principle 1: The Sun is the Primary Source of Energy for Earth's Climate System.	Core Idea ESS1: Earth's Place in the Universe • ESS1.C: The History of Planet Earth	Big Idea 1: Climate Change is a real and serious problem facing global society in the coming decades and centuries.
Essential Principle 2: Climate is regulated by complex interactions among components of the Earth system.	Core Idea ESS2: Earth's Systems • ESS2.A: Earth Materials and Systems • ESS2.C: The Roles of Water in Earth's Surface Processes • ESS2.D: Weather and Climate • ESS2.E: Biogeology	Big Idea 2: Climate change in recent decades is primarily caused by human activities, especially as related to energy use.
Essential Principle 3: Life on Earth depends on, is shaped by, and affects climate.	Core Idea ESS3: Earth and Human Activity • ESS3.A: Natural Resources • ESS3.B: Natural Hazards • ESS3.C: Human Impacts on Earth Systems • ESS3.D: Global Climate Change	Big Idea 3: Humans can take actions to reduce climate change and its impacts.
Essential Principle 4: Climate varies over space and time through both natural and man-made processes.	Core Idea ETS1: Engineering Design • ETS1.A: Defining and Delimiting an Engineering Problem • ETS1.B: Developing Possible Solutions • ETS1.C: Optimizing the Design Solution	Big Idea 4: To understand (deep) time and the scale of space, models and maps are necessary.

(continued)

TABLE 2.3 *(continued)*

Climate Literacy Principles (NOAA, 2024)	Next Generation Science Standards (NGSS, 2020)	Teacher-Friendly Guide to Climate Change (PRI, 2017)
Essential Principle 5: Our understanding of the climate system is improved through observations, theoretical studies, and modeling.	Core Idea ETS2: Links Among Engineering, Technology, Science, and Society • ETS2.A: Interdependence of Science, Engineering, and Technology • ETS2.B: Influence of Engineering, Technology, and Science on Society and the Natural World	Big Idea 5: The Earth is a system of complex systems.
Essential Principle 6: Human activities are impacting the climate system.		Overarching Question 1: How do we know what we know?
Essential Principle 7: Climate change will have consequences for the Earth system and human lives.		Overarching Question 2: How does what we know inform our decision-making?

Note. NOAA is part of the US Department of Commerce and their mission is to keep citizens informed about the changing environment using big data collection systems to monitor Earth's systems. The NGSS has several Earth and Space Science (ESS) standards as well as Engineering, Technology, and Application of Science (ETS) standards for educators to cover. *The Teacher-Friendly Guide to Climate Change* (TFGCC) synthesizes major ideas for educators to consider as they develop curricula around climate science.

Climate Change as a Socioscientific Issue

Climate change serves as the ideal vehicle to address the NGSS and scientific literacy, vice versa, the framework can be leveraged to advance much needed climate action through local classrooms. Which perspective resonates with you? How might that inform your approach? Looking at ethical issues related to the climate crisis, it's important to remember that climate change currently and will continue to impact people differently. Students need to understand how humans have and are the drivers for anthropogenic climate change, how they are directly impacted now and into the future, and how Earth's systems are already changing. We are already experiencing the impacts of climate change and we can support students as agents of change by providing the tools and skills needed to explain and take action on this devastating phenomenon. So how can we teach about climate change through the NGSS to develop scientifically literate students? Prior research tells us that teaching climate change as a **socioscientific issue** is key (Hestness, McDonald, Breslyn, McGinnis, & Mouza, 2014; Hestness, McGinnis, Riedinger, & Marbach-Ad, 2011; Holthuis, Lotan, Saltzman, Mastrandrea, & Wild, 2014; Matkins & Bell, 2007; Sadler, Chambers, & Zeidler, 2004).

Socioscientific issues (SSI) are complicated, open-ended, and controversial without straightforward solutions (Sadler et al., 2004). As an SSI, climate change is a politically controversial issue that students and the public often hear about in the media (ECCLPs, 2024). When the majority of teachers shy away from teaching about the climate crisis for various reasons, the media and Internet become primary sources of information for climate change (Caranto & Pitpitunge, 2015; Carter & Wiles, 2014; Hestness et al., 2014; Kolstø, 2001). It is important to note that students will come across climate change information regardless of whether or not teachers include or omit the content in the curriculum (Kolstø, 2001; Sadler et al., 2004). By being intentional with *how* we teach about the climate crisis, we can help students develop as 21st-century scientifically literate citizens capable of pushing back against disinformation campaigns or misleading information put forth by the media (Bunten & Dawson, 2014; Lambert & Bleicher, 2013).

Although the scientific data supporting climate change is not widely disputed, most educators and the public still doubt scientific consensus around the issue (Cook et al., 2013; Leiserowitz, Maibach, Roser-Renouf, Rosenthal, & Cutler, 2017; Somerville & Hassol, 2011). Rather than teaching the data that supports climate science, some teachers avoid or teach their personal opinion on the subject (Hestness et al., 2014; McCaffrey, 2015; Plutzer et al., 2016). This is problematic since the data supporting climate science is not disputed, but is so politically controversial that people think it is debatable. **Socioscientific issues include disagreements related to conflicting evaluations of the validity or trustworthiness of science-related claims.** Teaching climate change as an SSI allows for students to build their analytical skills through questioning, data analysis and interpretation, discourse, critical thinking, and more.

The SSI Framework encourages teachers to address social and ethical dimensions of climate change to support students as change agents. We are seeing an uprising across the world with young people leading protests in their cities demanding that governments take action to bend the curve. The environmental issues that students feel empowered to act upon and take more interest in are related to climate change's controversial nature and their right to learn the science and disproportionate impacts of climate change. An example of this is when students learn about disparate racial, social, and environmental injustices (directly resulting from the climate crisis) through community data analysis by zip code. Seeing the direct impact, students may also become highly interested to understand how they can influence local policy and ways to involve their community in taking action. They want change they can **see** and **feel** for their communities, and teachers can provide safe spaces and opportunities for students to hold these conservations and ideation around complex issues.

The consequence of not addressing the political interests, social values, or ethical implications underlying the climate crisis is that we *disempower* students (Hodson, 2003; Plutzer et al., 2016). Hodson (2003) argues that avoiding judgments in science is not possible since values are embedded in the science curriculum whether teachers recognize it or not. It is also important to note that values can also be promoted by content that is omitted from the classroom just as much as what is included. He further adds that the purpose of education should be to empower individuals to critically analyze society and values needed to sustain or change it (2003). Students need to ask what can be changed and how they can make those changes to address direct community impacts and ensure environmental sustainability. I invite you to complete Exhibit 2.2 to gauge your level of comfort and willingness to incorporate ethical dimensions related to the climate crisis.

Exhibit 2.2 Leaning into the Ethical Dimensions of Science Ed

Research reveals that teaching climate change as an SSI will empower your students to be agents of change in their own communities. It is important to gauge where you are in regards to addressing the ethical dimensions that relate to climate change to build student capacity.

Consider answering the following questions when you're ready:

☐ How often do you currently teach about social, racial, or political science issues in your class now? Why?

☐ Is there a possibility for interdisciplinary work with other subject matters on climate change (Social Science, Mathematics, or Language Arts subjects)? Is that a priority for you?

☐ How often do you include current science events directly related to climate change in your class? Why?

☐ How important is it to provide students with discourse opportunities (such as argumentation) on relevant scientific issues? Why?

☐ Can you imagine integrating opportunities for students to address ethical issues around climate change in the future (Some example issues include access to clean water, pollution, human health impacts, green technology, consumerism driving the economy, climate change policy, nuclear power, etc.)?

☐ How comfortable are you with teaching content that you might not have all the answers to? Does that determine whether or not you include it in your curriculum?

☐ Climate change is tentative in nature because there are no clear solutions and the data is still being collected showing the variety of impacts on Earth. How comfortable are you teaching about this content given that the information is constantly evolving?

☐ Scholars often argue that the educational system is about reinforcing the status quo (i.e. US education was never meant for historically marginalized and underrepresented minorities and only recently has been challenged). How much time and energy are you willing to invest to disrupt science education to provide accessible lessons to all students?

☐ What might be the purpose of science education? In what ways is your current instruction aligned to your vision? In what ways could it better align?

☐ Teachers are part of the education institution. Which statement resonates *most* with you and why? How might your thinking influence (or not) your teaching decisions?

    a. Systems are created and reinforced by people. As a teacher I make up this system, therefore I can change it.

    b. Systems are created and reinforced by people. I'm part of a larger system and have to abide by the rules and regulations in place.

Reflect on answers, and consider where you would like to shift your practices. Remember that students will be confronted by the ethical dimensions of climate change whether or not you address them in your class. How important is the climate crisis to you as it informs your curriculum design decisions? How will you help them use science to make informed decisions regarding those issues? How might students do science and engineering to mobilize on these issues?

Connect with the network to see what other teachers are doing to magnify their impact on education for climate action. See how others are addressing the ethical dimensions of the climate crisis at www.EmpoweredScienceTeachers.com (Book Resources → Chapter 2 → Discussion Board)

Deeper Dive into SSI to Inform Instruction

The SSI framework consists of three core components while also acknowledging the role of *classroom environments* (any norms and expectations established) and *peripheral influences* (external factors that might influence the outcome). The three core aspects presented by Presley et al. (2013) are *Design Elements, Learner Experiences,* and *Teacher Attributes.* Researchers argue that the SSI based framework supports scientific literacy by taking into account "real-life" scientific situations that may be influenced by other factors such as politics, social, or ethical issues. This perspective is also consistent with the Science and Engineering Practices of the NGSS.

Presley et al. (2013) emphasizes that each of the core aspects of the SSI framework identifies important features of teaching and learning. *Design Element* refers to creating instruction around compelling issues, presenting it first, scaffolding higher-order practices, and providing a culminating experience for students. The central issue must be compelling social issues with clear connections to science (such as climate change or evolution). The instruction is based on providing real-world contexts so that students will gain a deeper understanding of the science while developing skills needed to make decisions beyond the classroom. *Learner Experiences* focuses on further developing students' higher-order practices (reasoning, argumentation, decision-making, or positionality), confronting scientific theories, collecting and analyzing data, and considering other dimensions that may have an influence (social, racial, political, or economical). This core aspect also pushes students to confront ethical dimensions of SSI while applying the NOS principles in their analysis. *Teacher Attributes* look at how familiar the teacher is with SSI, sees teachers as learners, and considers their willingness to deal with uncertainties that may arise from addressing these open-ended controversial topics. Presley et al. (2013) stressed that successful SSI instruction relies heavily on teacher awareness of any social considerations related to the topic (e.g. to teach climate change effectively you have to consider the politically controversial nature of the topic that influences students' current understandings).

Connection to the Nature of Science

Integrating controversial SSI such as climate change into your science curriculum is an opportunity for you to teach NOS principles and NGSS. Recall that the NOS calls for students to examine a topic's empirical evidence, social and cultural embeddedness, and tentative nature (Sadler et al., 2004). These major components of the SSI framework push students to utilize major NOS themes to formulate ideas regarding climate change on evidence. Students' are likely to accept or take action on controversial SSI after understanding the nature of scientific knowledge (Carter & Wiles, 2014; Khishfe & Lederman, 2006; Kolstø, 2001; Sadler, Barab, & Scott, 2007). Although this approach can effectively strengthen scientific literacy, evidence reveals that teachers need more learning opportunities and implementation support that effectively integrate the NOS through SSI to help students become informed decision-makers (Kolstø, 2001; Lee, Chang, Choi, Kim, & Zeidler, 2012; Matkins & Bell, 2007; Sadler et al., 2004, 2007).

NOS is necessary for students to successfully confront politically controversial topics they will encounter in their daily lives (Caranto & Pitpitunge, 2015; Hodson, 2003; Lambert & Bleicher, 2013; Shea, Mouza, & Drewes, 2016). Teaching students science as a SSI is necessary to help students align their thinking with the scientific community and to become informed and engaged citizens (Sadler et al., 2007). Individuals who take action on SSI have a deep and personal understanding of the issue, and feel they have a personal investment in addressing or solving the problems (Hodson, 2003; Kolstø, 2001). Scientific literacy needs to equip students with knowledge and abilities to apply NOS principles to effectively mobilize for environmental and climate justice.

Moving Forward with Confidence

Looking at the available resources, it is important to acknowledge that there is no curriculum that will work for every science teacher universally. Science curricula should be tailored to the needs of students, their community, and responsive to their lived realities. It also needs to be respectful of teachers as pedagogical leaders that know their schools and students, give the freedom to approach teaching through different phenomena, and allow for multiple ways to gauge students' progress. If we are going to empower students to take action on the climate crisis, we need to teach the science of climate change and provide opportunities to engage in argumentation from evidence about the social and ethical implications. Consider connecting with your local environmental community partner or higher education institution to unveil relevant challenges, solutions, and innovations. See Exhibit 2.3 for an example of such collaboration and how it can enhance lesson design around climate and environmental topics.

Exhibit 2.3 Collaboration is Key for Cultural Relevance

This lesson was created in collaboration with Dr. Shelley Brooks (UC Davis), who supported me in learning and thinking about the integration of social science in meaningful ways. This multi-day lesson was designed to build students' capacity and complex critical thinking skills by learning about the urban Heat Island Effect, importance of tree canopy coverage, redlining and historical inequities. Access the full lesson, slides, and worksheets at bit https://www.bit.ly/3VhvXHt

Total Time

Fifty-five-minute lessons (3 class periods with the option to extend across several classes if desired)

Learning Outcomes

Students will learn about the historical underpinnings for inequities of redlining and its impact on human health in specific communities by understanding tree canopies and green spaces. In addition, students will apply their knowledge by creating an infographic to use art that inspires action to share what they learned with others.

Free Interactive Tools

◆ Park Score Ranking—www.tpl.org/parkscore
◆ Park Access Interactive—bit.ly/3TCYOVs
◆ CalEnviro Interactive—bit.ly/4a8ILUU
◆ Adobe Spark—adobe.ly/4abNepu

Multi-day Lesson Plan Scope and Sequence—*Teacher Guidance: Please use teaching slides with students and read the notes section for guidance, teacher tips, and additional learning resources to extend the experience if desired.*

Day 1—The Importance of City Parks

◆ Part 1 (Draw out Students' Funds of Knowledge): Engage students in a whole group discourse by using images where students recognize areas of tree canopy/green spaces compared to areas with less trees/green spaces. Have students share about their own neighborhoods in comparison and if trees/green spaces matter.
◆ Part 2 (Park Score Index Map): Students will explore and use the *Park Score Ranking map* and analyze city data including the location, sizes, and access to city parks. Use *student worksheet #1* to guide students' observations and questions.
◆ Part 3 (Park Access Interactive Tool): Students will further their knowledge by exploring the *Park Access Tool* looking at local community data that highlights the amount of residents living in proximity to green spaces and parks (leaning into access to green spaces). Students will visualize and analyze the distribution of green spaces and parks and map of historically marginalized communities in low-income neighborhoods.

◆ Part 4 (Engage Students in Discourse About Redlining): Debrief with students for sense-making. Engage students in a small or whole group discourse by using images where students look for patterns between redlining markers and human health in their surrounding communities.

Day 2—Trees and the Urban Heat Island Effect

◆ Part 1 (Draw out Students' Funds of Knowledge): Engage students in small or whole group discourse about the benefits that trees provide humans by displaying images that show temperature differences in different areas and connection to tree canopy coverage. The higher temperature differences in urban areas is commonly known as the "Heat Island Effect."
◆ Part 2 (CalEnviro Tool): Students will explore the *Cal Enviro Tool* to map out air quality in local and surrounding communities. Use *student worksheet* #2 to guide students' observations and questions.
◆ Part 3 (Engage Students in Discourse of Solutions for a Complex Ecosystem): Debrief with students for sense-making. Engage students in discourse around central questions as they consider meaningful action.

Day 3—Use Art to Inspire Action

◆ Activate Student Agency: Students will create an infographic choosing any aspect that they learned about to address environmental justice exacerbated by the climate crisis that this lesson highlights. Students will also think about and plan for a call to action using their infographic. Consider having students use *Adobe Spark* to first make a moodboard to brainstorm their design using art for inspiration. Lastly, students can share the infographic with family, friends, other teachers, and/or the community and attach their own QR code or URL link to engage their audience in learning more.

Collaborating with Dr. Brooks helped me to understand the importance of historical injustices and practices of erasure that highly impact marginalized communities at the center of these intersectional issues. Her guidance and expertise taught me about redlining tied to human health impacts (when looking at excess ER visits in comparison to the HOLC maps). My partnerships with local aquarium outdoor educators also taught me about the importance of native Indigenous plants, failed environmental efforts in Los Angeles that resulted in significant changes to the local salt marsh ecosystem, and ways that students can think more critically about seemingly simple solutions in terms of trade-offs, risks, and potential benefits. This lesson plan went beyond the science, and gave students opportunities to build engineering habits of mind and critical thinking skills to real and relevant complex issues.

 *Connect with the network to see how other teachers are connecting with environmental community partners or higher education faculty and staff to enhance their lesson plans at www.EmpoweredScienceTeachers .com (Book Resources → Chapter 2 → Discussion Board)*

When you are ready for implementation support, the next chapters serve as learning, planning, and teaching guides to help you integrate climate science through NGSS. Continue to track your professional growth by re-examining your beliefs about teaching and learning using the three dimensions of the SSI Framework.

 *Refer back to the Book Introduction where you initially tracked your professional growth. Reflect on how your beliefs might have changed and why that might be. How might your responses unveil more about your teaching disposition?*

Tracking Your Professional Progress

Student Learning Experiences in Your Class	On a scale of 1 (extremely disagree) to 5 (extremely agree), to what extent do I agree or disagree with the following statements? Why?
Today's Date:	
Students often discuss policies related to science.	
Students often collect and analyze current data or information.	
Students often discuss ethical issues related to science.	
Students often co-construct knowledge with me during lessons.	
Students often drive the instruction in my class as capable contributors and doers of science.	
Students are often positioned as current scientists or engineers.	
Students often learn about Indigenous traditional ecological knowledge.	
Curriculum Design Elements	On a scale of 1 (extremely disagree) to 5 (extremely agree), to what extent do I agree or disagree with the following statements? Why?
Today's Date:	
I often build lessons/units around anchoring or investigative phenomena.	
I often present climate or environmental issues at the start of each unit or lesson.	
Students often engage in argumentation and making claims based on evidence.	

Curriculum Design Elements	On a scale of 1 (extremely disagree) to 5 (extremely agree), to what extent do I agree or disagree with the following statements? Why?
Students often engage in meaningful discourse opportunities.	
My lessons/units are often centered around real-world issues that are directly related to students' lives or community.	
Students often engage with the Nature of Science principles.	
Students often connect with local or relevant field experts and researchers.	

My Teacher Attributes	On a scale of 1 (extremely disagree) to 5 (extremely agree), to what extent do I agree or disagree with the following statements? Why?
Today's Date:	
I have to know everything about a particular topic before teaching it.	
I am extremely confident in my knowledge of climate science.	
I feel comfortable admitting to students when I don't know the answer to their question.	
I am comfortable teaching about open-ended issues where I cannot predict student responses.	
I am not the only source of knowledge on climate and the environment for students.	
I often experience imposter syndrome when teaching, even for topics that I have strong expertise in.	
I often reflect on my pedagogical decisions to improve my practice.	

Collective Voices for Climate Change Education

Tamara Wallace, Assistant Director of Energy, Sustainability, and Transportation California State University, Office of the Chancellor

I believe public higher education is a collective space that drives opportunities for innovation, creativity, and learning. It is an access point for equity and mobility for our society. Universities serve as a critical community pillar in times of crisis. College campuses are an essential workforce development engine for the emerging generation of our society as our future teachers, nurses, engineers, business developers, artists, historians, politicians, and community leaders. We, in higher education, have an inherent obligation to our communities we serve to uphold the values of environmentalism and conservation so as to instill these ideals upon our future leaders, educators, and parents. As a mother, I see that my own young children and their fellow generation will be directly impacted by the apparent and urgent paradigm shift currently taking place in education that aims to adequately prepare them intellectually and emotionally for the realities of the planet with a changing climate.

Karolyn Burns, Education and Curriculum Manager at The CLEO Institute

I am inspired by seeing students grasp fundamental media literacy principles while exploring climate and environmental issues. It's awesome to see them do research when consuming news, and discern and challenge misinformation. There is no doubt that this essential skill extends beyond the classroom, becoming invaluable in navigating the constant influx of information in both academic and real-world contexts. Providing students with the ability to discern credible sources is a gift that will undoubtedly empower them throughout their educational journey and into their future careers. The fact that they have such a strong desire to take action, as well as the ability to assess information accurately, allows them to lead and protect their communities against climate change.

Additional Teacher Resources

Access the NGSS standards by topic—
Bit.ly/NGSSTOPIC
Explore climate change resources with NSTA—
NSTA.org/topics/climate-change
Free download of The Teacher-Friendly Guide to Climate
Change—Bit.ly/PRITFG
Learn more about achieving climate and environmental
stability—Bit.ly/BigCCReport
Learn more about climate teaching resources with CLEAN—
Bit.ly/CLEANGCC
MADE CLEAR's outline of climate science in NGSS—
Bit.ly/MADECLEARGCC
Read more about NOAA's Climate Literacy Principles—
Bit.ly/CCNOAA
Read more about The Socioscientific Issues Framework—
Bit.ly/SSIFRAME

Note

1   Lyla June's remarks were made at the 2024 RegenIntel Fellows workshop hosted online by Regenerative Intelligence.

References

Bowman, T., Morrison, D. (2020) An ACE National Strategic Planning Framework for the United States [Online]. Created in collaborative reflection with the U.S. ACE Community. Available at http://aceframework.us.

Bunten, R., & Dawson, V. (2014). Teaching climate change science in senior secondary school: Issues, barriers and opportunities. *Teaching Science*, 60(1), 10.

Caranto, B. F., & Pitpitunge, A. D. (2015). Students' knowledge on climate change: Implications on interdisciplinary learning. In Biology Education and Research in a Changing Planet: Selected Papers from the 25th Biennial Asian Association for *Biology Education Conference* (pp. 21–30). Springer Singapore.

Carter, B. E., & Wiles, J. R. (2014). Scientific consensus and social controversy: Exploring relationships between students' conceptions of the nature of science, biological evolution, and global climate change. *Evolution: Education and Outreach*, 7(1), 6.

Cook, J., Nuccitelli, D., Green, S. A., Richardson, M., Winkler, B., Painting, R., ... & Skuce, A. (2013). Quantifying the consensus on anthropogenic global warming in the scientific literature. *Environmental Research Letters*, 8(2), 024024.

ECCLPs. (2024). ECCLPs HS student survey report. ECCLPs. https://ecclps.net/hs-ss-report

Hestness, E., McDonald, R. C., Breslyn, W., McGinnis, J. R., & Mouza, C. (2014). Science teacher professional development in climate change education informed by the next generation science standards. Journal of Geoscience Education, 62(3), 319–329.

Hestness, E., Randy McGinnis, J., Riedinger, K., & Marbach-Ad, G. (2011). A study of teacher candidates' experiences investigating global climate change within an elementary science methods course. Journal of Science Teacher Education, 22(4), 351–369.

Hodson, D. (2003). Time for action: Science education for an alternative future. International Journal of Science Education, 25(6), 645–670.

Holthuis, N., Lotan, R., Saltzman, J., Mastrandrea, M., & Wild, A. (2014). Supporting and understanding students' epistemological discourse about climate change. Journal of Geoscience Education, 62(3), 374–387.

Khishfe, R., & Lederman, N. (2006). Teaching nature of science within a controversial topic: Integrated versus nonintegrated. Journal of Research in Science Teaching, 43(4), 395–418.

Kolstø, S. D. (2001). Scientific literacy for citizenship: Tools for dealing with the science dimension of controversial socioscientific issues. Science Education, 85(3), 291–310.

Lambert, J. L., & Bleicher, R. E. (2013). Climate change in the preservice teacher's mind. Journal of Science Teacher Education, 24(6), 999–1022.

Lee, H., Chang, H., Choi, K., Kim, S. W., & Zeidler, D. L. (2012). Developing character and values for global citizens: Analysis of pre-service science teachers' moral reasoning on socioscientific issues. International Journal of Science Education, 34(6), 925–953.

Leiserowitz, A., Maibach, E., Roser-Renouf, C., Rosenthal, S., & Cutler, M. (2017). Climate change in the American mind: May 2017. Yale University and George Mason University. New Haven, CT: Yale Program on Climate Change Communication.

Matkins, J. J., & Bell, R. L. (2007). Awakening the scientist inside: Global climate change and the nature of science in an elementary science methods course. Journal of Science Teacher Education, 18(2), 137–163.

McCaffrey, M. S. (2015). Climate smart & energy wise: Advancing science literacy, knowledge, and know-how. Thousand Oaks, CA: Corwin.

NGSS Lead States. (2013). Next generation science standards: For states, by states. Washington, DC: The National Academies Press.

Paas, L., & Goodman, D. (2016). Action for climate empowerment: Guidelines for accelerating solutions through education, training and public awareness. Paris and Bonn: United Nations Educational, Scientific and Cultural Organization and the Secretariat of the United Nations Convention on Climate Change, p. 2. Retrieved from https://unfccc.int/sites/default/files/action_for_climate_empowerment_guidelines.pdf

Plutzer, E., McCaffrey, M., Hannah, A. L., Rosenau, J., Berbeco, M., & Reid, A. H. (2016). Climate confusion among US teachers. Science, 351(6274), 664–665.

Presley, M. L., Sickel, A. J., Muslu, N., Merle-Johnson, D., Witzig, S. B., Izci, K., & Sadler, T. D. (2013). A framework for socio-scientific issues based education. Science Educator, 22(1), 26.

Sadler, T. D., Barab, S. A., & Scott, B. (2007). What do students gain by engaging in socioscientific inquiry? Research in Science Education, 37(4), 371–391.

Sadler, T. D., Chambers, F. W., & Zeidler, D. L. (2004). Student conceptualizations of the nature of science in response to a socioscientific issue. International Journal of Science Education, 26(4), 387–409.

Shea, N. A., Mouza, C., & Drewes, A. (2016). Climate change professional development: Design, implementation, and initial outcomes on teacher learning, practice, and student beliefs. Journal of Science Teacher Education, 27(3), 235–258.

Somerville, R. C., & Hassol, S. J. (2011). The science of climate change. Physics Today, 64(10), 48.

The Essential Principles of Climate Literacy: NOAA Climate.gov. (2009). Retrieved October 15, 2020, from https://www.climate.gov/teaching/essential-principles-climate-literacy/essential-principles-climate-literacy

Zabel, I. H. H., Duggan-Haas, D., and Ross, R. M., eds. (2017). *The teacher-friendly guide to climate change*. Ithaca, NY: Paleontological Research Institution, 284 pp.

Part 2

Developing Scientific Literacy Using Climate Science

3

Climate Change Is Complex, Where Do I Start?

Read this when:

- ♦ *You're ready to learn the fundamentals of climate science.*
- ♦ *You need guidance on what to teach regarding climate change.*
- ♦ *You want to know research-based approaches to successfully teach about climate science.*

Climate Change Education at Scale

The more that I connect with teachers and leaders across the world, the more I have come to understand the importance of building coalitions and spaces for educators and students to connect over climate change education and action. We can learn so much from one another, share key approaches to teaching about climate and environmental topics, and continue to build students' emotional resilience and collaboration skills through virtual connections with peers aiming to tackle local challenges. Given the large scale of the problem, it is understandable that teachers might not know exactly what to teach about. Where should you start integrating climate science? How could you teach it in ways that empower young people, while also fulfilling

DOI: 10.4324/9781003478584-6

the NGSS? What approaches are working for other educators and how can we learn more? In a project I led through ECCLPs, leaders across subject areas and grades were able to identify gaps and opportunities across teaching and learning for climate education that are noteworthy when thinking about approaches at scale. Review Table 3.1 for the paradigm shifts needed to support 21st-century climate change education efforts.

TABLE 3.1 Key Shifts in Teaching and Learning for Climate Literacy

From *Current Broadly Implemented Approaches*	Toward *More Holistic and Nuanced Approaches to Activate Student Agency (ECCLPs, 2024)*
1. Positioning climate change solutions as reductionistic, highly technical, far-removed from those it impacts, and universally effective across all communities and settings.	1. Framing climate change solutions as dynamic, innovative, interconnected, and sensitive to the different, layered and overlapping vulnerabilities experienced by different communities, necessitating everyone's involvement in generating diverse and responsive solutions, and underscoring the crucial role of education in fostering understanding and engagement about this socioscientific issues, its causes and solutions.
2. Little to no focus on the impacts of climate change on social inequity and injustice, and vice versa.	2. A holistic approach that acknowledges the interplay between social justice, environmental justice, racial justice, and indigenous sovereignty, emphasizing the unequal distribution of climate impacts and accountability. It shifts focus from individual blame to systemic responsibility, recognizing the need for collective action to address these intertwined issues.
3. Positions climate-related expertise as only coming from a limited number of scientific disciplines.	3. Recognizing and integrating expertise and knowledge from multiple sources, including western knowledge (science, math, history, social science, etc.), Indigenous knowledge, local community knowledge, etc., while emphasizing humility, lifelong learning, and inter- and transdisciplinary education from early grades.

From *Current Broadly Implemented Approaches*	Toward *More Holistic and Nuanced Approaches to Activate Student Agency (ECCLPs, 2024)*
4. A focus on abstract content knowledge, rather than skills for application or action.	4. Empowering students with essential 21st century skills, competencies, and dispositions necessary for active civic engagement, and integrating social emotional learning and empathy to address tough emotions about climate change and to foster hope, agency, and adaptation.
5. Incoherent and inconsistent pedagogical practices that are often influenced by political pressure while ignoring the multigenerational and institutional wisdom of Indigenous peoples, as well as community practices, deeply held beliefs, and values that influence actions.	5. Pedagogical practices and approaches that are continuously evolving to reflect research-driven approaches, Indigenous knowledge and other ways of knowing to be locally, culturally, and linguistically relevant and action-oriented.
6. Engaging with students as passive learners or "empty buckets" that must be filled by a teacher with expertise.	6. Empowering students as active partners and future stewards in driving climate solutions, fostering youth agency. This includes providing opportunities for action civics, enabling students to translate their knowledge into tangible actions that address societal challenges and promote positive change.

Credit: Environmental and Climate Change Literacy Projects (ECCLPs), 2024.

Although the resources above primarily address effective and necessary approaches to teaching climate change, where can you begin to jump into the fundamentals of climate science? To provide guidance, I draw upon the research of highly credible national and local educational agencies that focus on climate and environmental literacy (such as the National Aeronautics and Space Administration (NASA), National Oceanic and Atmospheric Administration (NOAA), National Center for Science Education (NCSE), Yale Program on Climate Change Communication (YPCCC), and Project Drawdown to name a few). This following section can seem technical and very detailed, but it will provide teachers with the fundamentals of climate science to develop a shared language and approach.

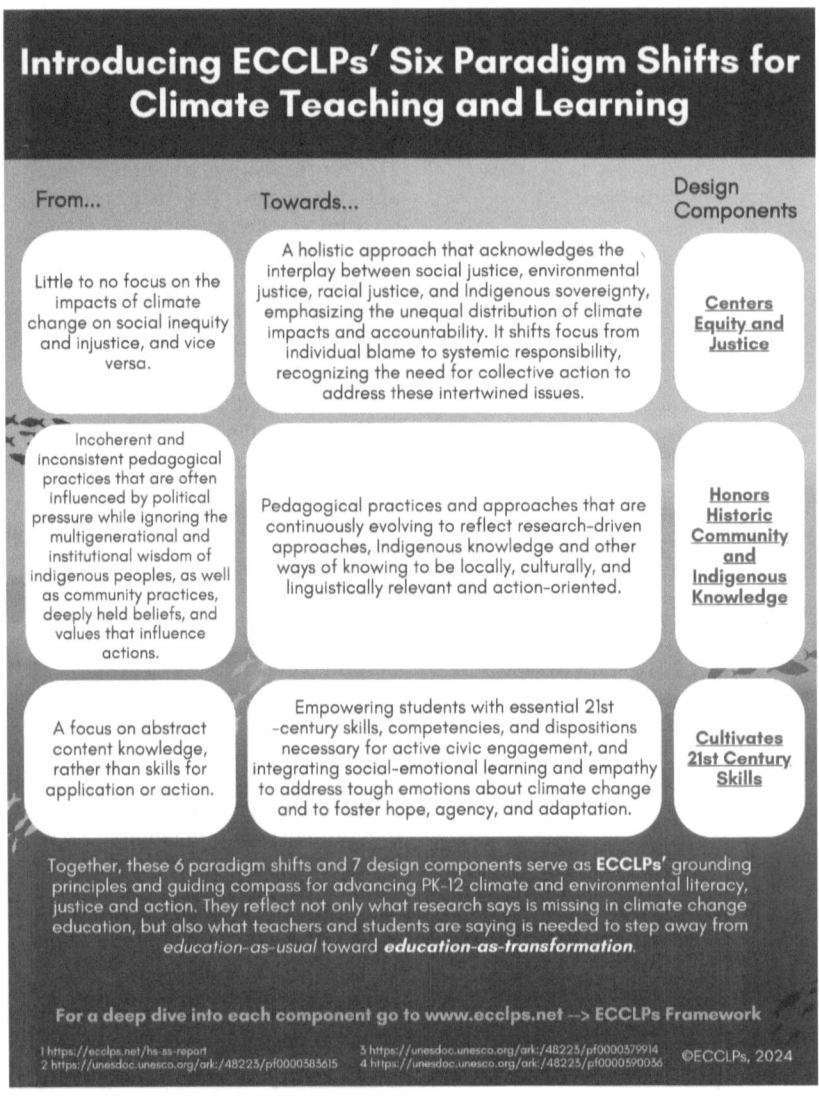

Image Credit: Environmental and Climate Change Literacy Projects (ECCLPs), 2024

Public Attitudes Regarding Climate Change

According to ongoing studies by Yale and George Mason University that began in 2008, there are six distinct groups that categorize Americans' beliefs, attitudes, policy preference, and

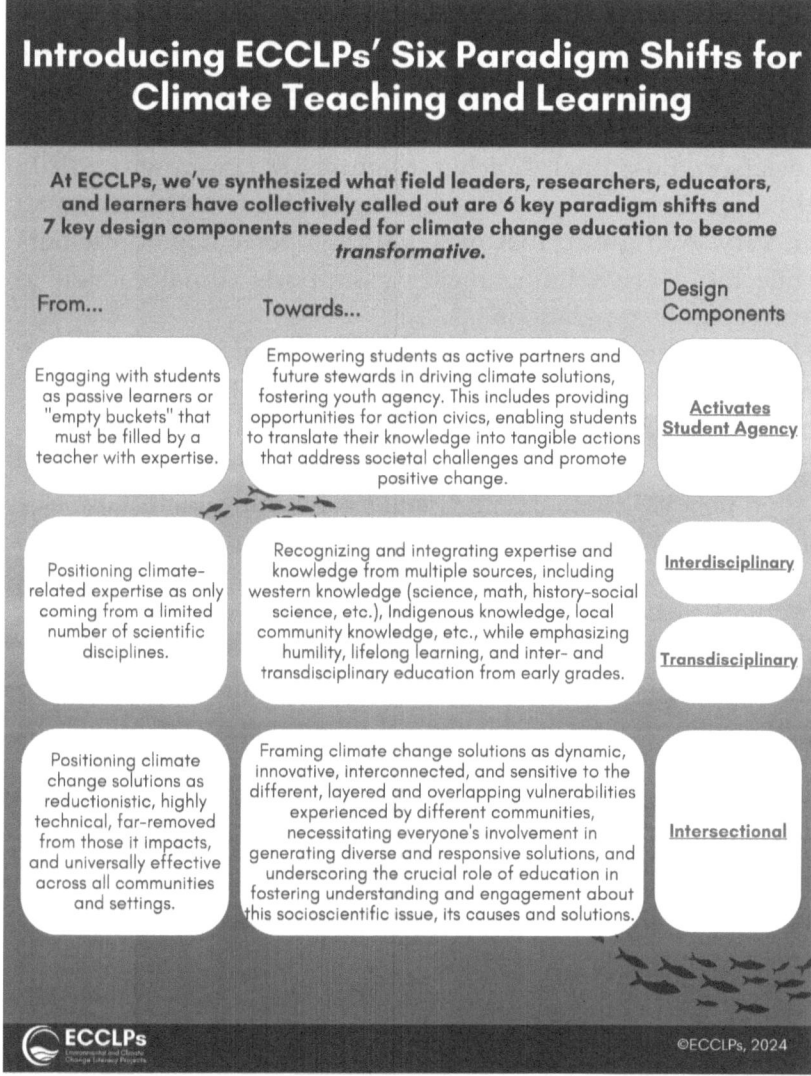

Introducing ECCLPs' Six Paradigm Shifts for Climate Teaching and Learning

At ECCLPs, we've synthesized what field leaders, researchers, educators, and learners have collectively called out are 6 key paradigm shifts and 7 key design components needed for climate change education to become *transformative*.

From...	Towards...	Design Components
Engaging with students as passive learners or "empty buckets" that must be filled by a teacher with expertise.	Empowering students as active partners and future stewards in driving climate solutions, fostering youth agency. This includes providing opportunities for action civics, enabling students to translate their knowledge into tangible actions that address societal challenges and promote positive change.	**Activates Student Agency**
Positioning climate-related expertise as only coming from a limited number of scientific disciplines.	Recognizing and integrating expertise and knowledge from multiple sources, including western knowledge (science, math, history-social science, etc.), Indigenous knowledge, local community knowledge, etc., while emphasizing humility, lifelong learning, and inter- and transdisciplinary education from early grades.	**Interdisciplinary** **Transdisciplinary**
Positioning climate change solutions as reductionistic, highly technical, far-removed from those it impacts, and universally effective across all communities and settings.	Framing climate change solutions as dynamic, innovative, interconnected, and sensitive to the different, layered and overlapping vulnerabilities experienced by different communities, necessitating everyone's involvement in generating diverse and responsive solutions, and underscoring the crucial role of education in fostering understanding and engagement about this socioscientific issue, its causes and solutions.	**Intersectional**

ECCLPs

©ECCLPs, 2024

Image Credit: Environmental and Climate Change Literacy Projects (ECCLPs), 2024

behavior regarding climate change. Known as the *Six Americas*, these researchers discovered six audiences that respond to climate change information differently. The Six Americas are classified as *Alarmed, Concerned, Cautious, Disengaged, Doubtful* or *Dismissive* (Howe, Mildenberger, Marlon, & Leiserowitz, 2015). In 2020, the *Alarmed* and the *Concerned* constituted 54%

of the population while 25% of the population held low beliefs about the climate crisis (see Figure 3.1). The year 2020 marks the first time that Alarmists are now the largest group in the Six Americas, and has tripled in size since 2014 (Goldberg, Gustafson, Rosenthal, Kotcher, Maibach, & Leiserowitz, 2020). Although the media portrays dissonance, it's really only a small but very loud group that makes up the percentage of deniers while the overwhelming majority supports climate action to protect future generations.

In 2020, the YPCCC also found that 72% of Americans think that climate change *is* happening (see Figure 3.2), but only 57% understand that it's primarily caused by human activities. The public opinion mapping tool also revealed that 43% of the nation believes that climate change is harming them now (see Figure 3.3), while 71% believe it will harm future generations (see Figure 3.4). Lastly, 75% of Americans support regulating CO_2 as a pollutant, and 86% want to fund more research on renewable energy sources despite conflicting media messaging. What's clear is that the data reveals many differing views and opinions about climate change and what approaches should be taken. Understanding the current climate is essential because among that population sample, are science educators who are now responsible for teaching climate science under the new

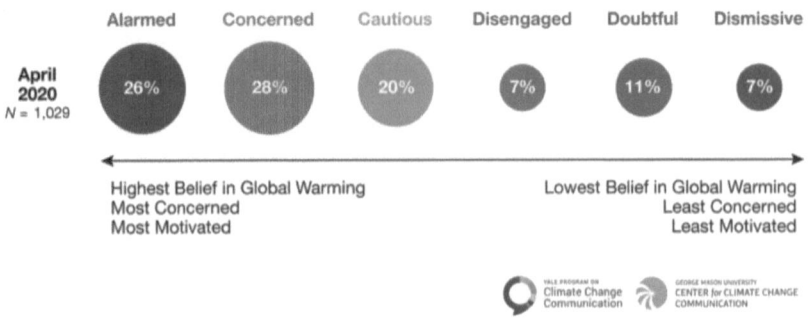

FIGURE 3.1 Yale's Report on the Six Americas
Image Credit: Maibach, E. W., Leiserowitz, A., Roser-Renouf, C., & Mertz, C. K. (2011). https://climatecommunication.yale.edu/about/projects/global-warmings-six-americas/

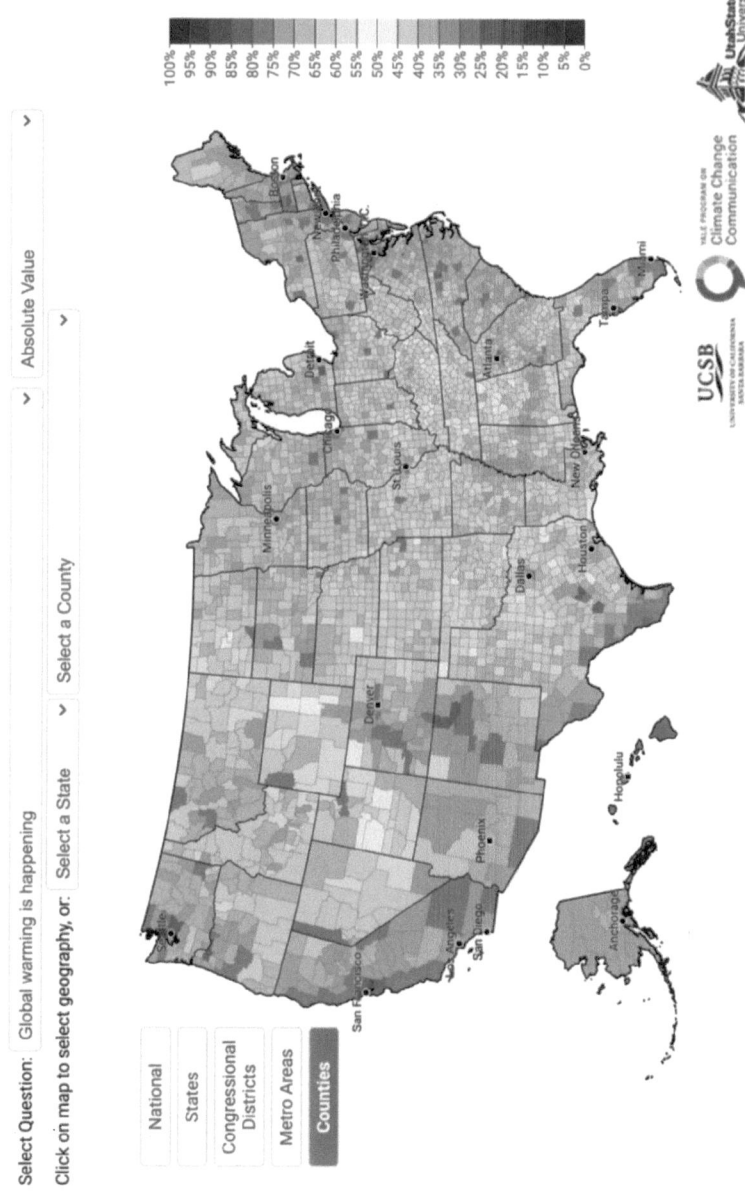

FIGURE 3.2 Yale Public Opinion Tool on Climate Change

Image Credit: Howe, P., Mildenberger, M., Marlon, J., & Leiserowitz, A. (2015) https://climatecommunication.yale.edu/about/projects/global-warmings-six-americas/

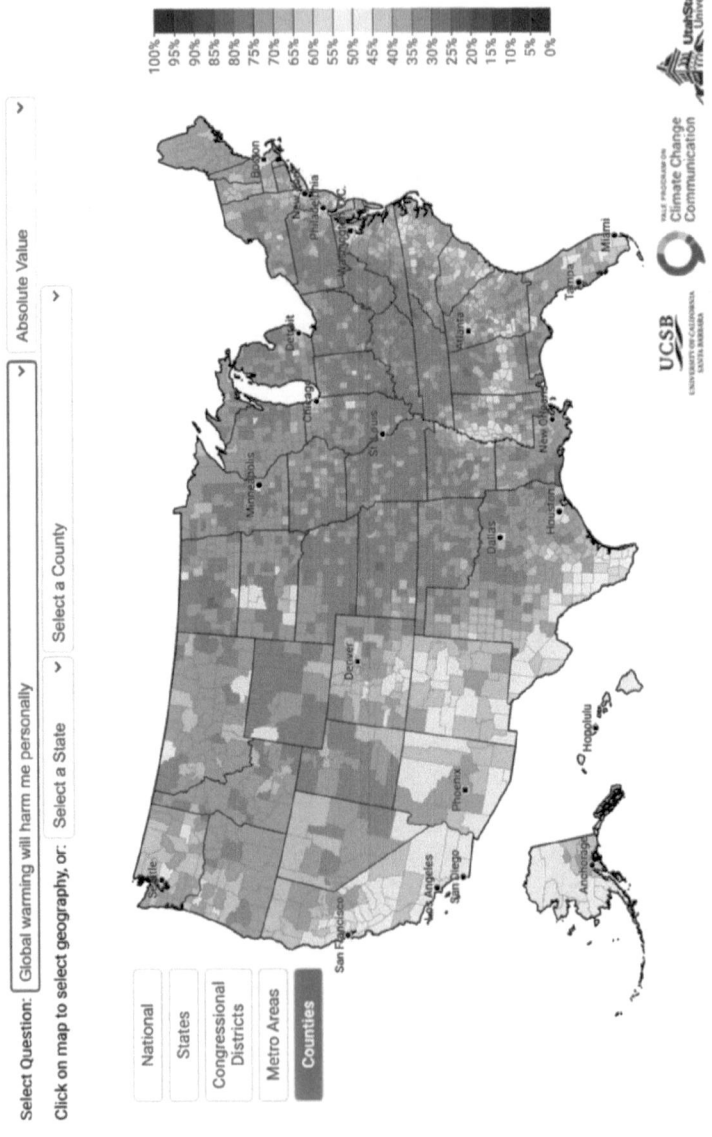

FIGURE 3.3 Yale Public Opinion Tool on Current Climate Impacts

Image Credit: Howe, P., Mildenberger, M., Marlon, J., & Leiserowitz, A. (2015) https://climatecommunication.yale.edu/about/projects/global-warmings-six-americas/

Estimated % of adults who think global warming will harm future generations (71%), 2020

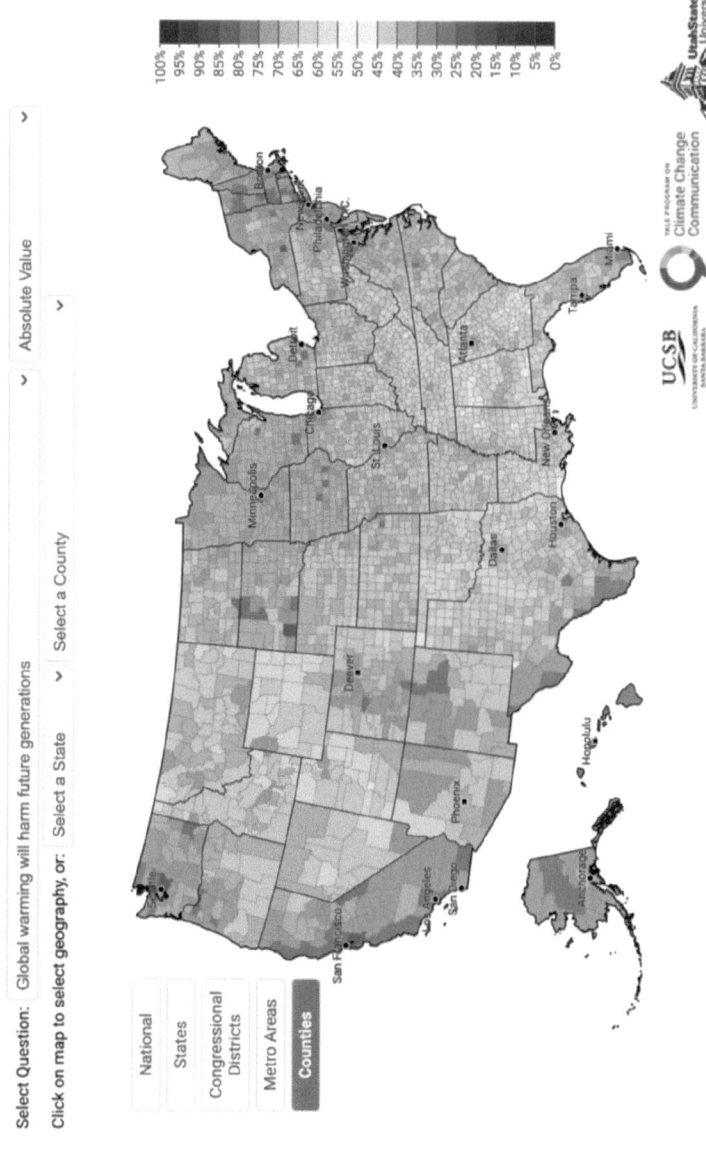

FIGURE 3.4 Yale Public Opinion Tool on Future Impacts of Climate Change
Image Credit: Howe, P., Mildenberger, M., Marlon, J., & Leiserowitz, A. (2015) https://climatecommunication.yale.edu/about/projects/global-warmings-six-americas/

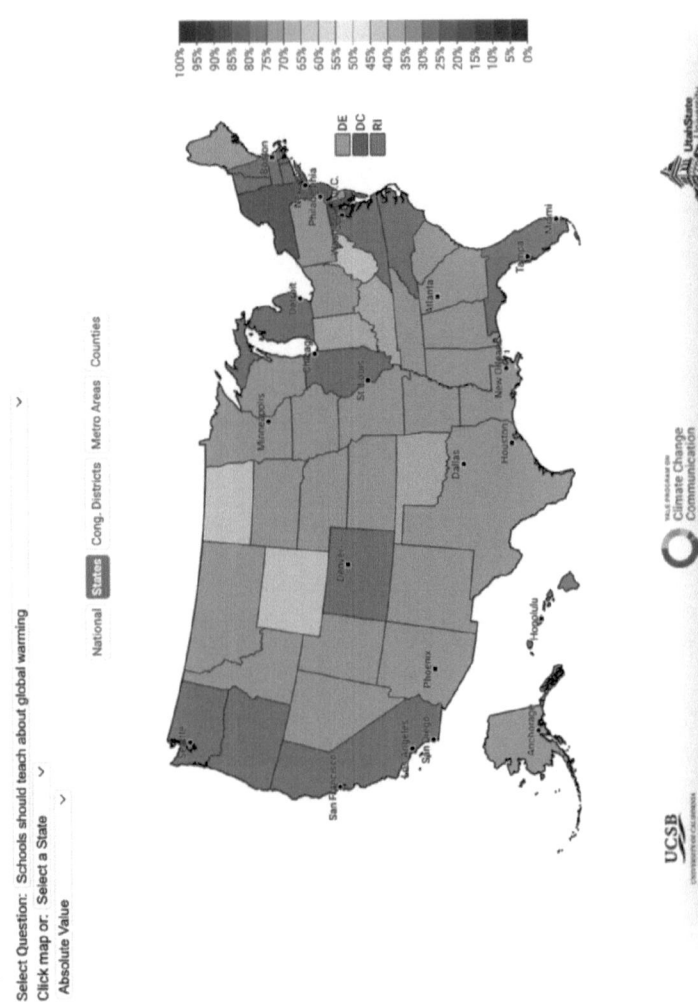

FIGURE 3.5 Yale Public Opinion Tool on Role of Schools

Image Credit: Howe, P., Mildenberger, M., Marlon, J., & Leiserowitz, A. (2015) https://climatecommunication.yale.edu/about/projects/global-warmings-six-americas/

framework. It's important to know that the public at large is extremely supportive of students learning about climate science in schools despite what the media portrays.

 *Consider taking and administering the Yale Six Americas Super Short Survey (SASSY) to see which audience you, your students, and their families identify with. Go to www. Bit.ly/YALEsassy and answer four short questions to get instant results and consider comparing class data with national averages.*

In looking at teacher attitudes regarding climate change, recall the 2016 study that revealed much confusion among science educators across the United States (Plutzer et al., 2016). The survey included 1,500 public secondary science educators and found that the median teacher allocates only 1–2 hours on climate change—covering nowhere near the depth of knowledge demanded by the NGSS. Roughly 52% of the teachers know the scientific consensus, and 68% of surveyed teachers understand that climate change is caused by human activities (Plutzer et al., 2016). Furthermore, the authors noted that nearly 50% of science educators did not learn climate science during their under-graduate career or teacher preparation programs. Therefore, more educational opportunities are needed to educate teachers to effectively integrate climate science into their curriculum (Hestness et al., 2014; Lambert & Bleicher, 2013; Liu, Roehrig, Bhattacharya, & Varma, 2015; Plutzer et al., 2016; Sadler et al., 2004).

When looking at 2023 public opinions on whose responsi-bility it should be to educate students about climate change, **YPCCC reveals that 75% of the public believe it needs to happen in schools** (Howe et al., 2015). Figure 3.5 shows nation-wide results of how people across all states agree that schools need to teach about climate change causes, consequences, and solutions. Teachers need educational support to teach the science of climate change, and it's helpful knowing that the general public supports and expects teachers to take on this role contrary to the conflicting messages broadcasted by mass media.

 *As you reflect on this information, consider what your learning needs are regarding climate science and how you would potentially work with colleagues to increase capacity for climate change education at your school.*

Identifying Your State of Mind

Your Current State

With a better understanding of public attitudes regarding the role of education and climate change, you can shift your focus to identifying learning goals you're ready to take on. We will refer to this state of mind as your **current state**. Acknowledging your current state is important because it means that you have reflected on your current theories of action, and recognize what is within your control as well as any potential challenges. The following activities in Exhibit 3.1 help to illustrate the possibility of change that can also be used with your students.

Exhibit 3.1 Activities to Illustrate Change

There are several activities that you can use to illustrate that change is possible. Consider taking some of these activities back to the class when you integrate climate change content to position students as capable agents of change. *Note: There are many versions of these activities online to explore.*

Arms Crossed Activity

1. Ask students to stand up and cross their arms in front of them as though they are waiting for a friend to come out of class during a passing period.
2. Ask them to hold their arms there and take note of whether the left or right arm is on the top.
3. Ask them to release the hold, shake it off. Then cross again naturally.

4. Ask students to share how that felt to just cross your arms (i.e. How did that feel? How easy was it to do that? Did you have to think about which arm goes on top or on the bottom? How do you think you learned to do this?)

5. Now, ask them to release the hold, shake it off. Then cross again but this time, ask them to purposely put the opposite arm on the top.

6. Ask students to share how that felt (i.e. How did that feel? How easy was it to do that? Did you have to think about which arm goes on top or on the bottom? How do you think you learned to do this?)

7. As they stand there awkwardly holding this pose that seems somewhat unnatural to them, facilitate a quick discussion about the possibility of change. Although it was weird to push students to control which arm goes on top, they did it. That is the point. Climate change is exacerbated because we take for granted that our everyday activities have a larger impact and although it would be "awkward," "weird," "uncomfortable," or even "difficult" to change our actions, it is important to note that it is **possible**. With enough practice over time, it would become second nature to do things a different way.

Sustainability Activity

1. Hold up two pens for students to examine (make sure one pen is a disposable pen and the other is a fancier looking pen).

2. Let students know that one pen costs about 10 cents and the other is a rare $100 collectable pen that you can buy ink cartridges for (it's okay to embellish because they will soon understand your point).

3. Ask them to explain to an elbow partner which pen would be considered more "sustainable" and "why they think that."

4. After facilitating this conversation and hearing from students or taking a class vote, have a discussion about sustainability and recognizing our collective relationship with material items.

5. The disposable pen may seem like the obvious choice to make as unsustainable because it's cheap and meant to be thrown away. It's important to be explicit about this and to also share that with these pens, you use them daily and rely on them for grading.

6. Most students would automatically say that the fancier pen is more sustainable because of its features. Explain that the fancier pen, however, was a gift given to you that is used once every few years when writing important letters. The pen is made from rare metals and the ink cartridge is expensive to replace. Ask students to rethink "sustainability" when relationships are considered.

7. The problem most students might not realize is that even if the fancier, more expensive pen is more sustainable (because you won't throw it away and plan to reuse it forever), we choose not to use it very often and our relationship with that item also contributes to making it unsustainable. If we're constantly opting for cheaper disposable options because they're more affordable at the moment, we have a comfortable and reliant relationship with those items, that is a problem. We need to change our relationship and thinking with the items we rely on in order to push ourselves to change.

*Note—This activity also works well with water bottles and canteens as well.

 *Share your teaching experiences with the network at www.EmpoweredScienceTeachers.com (Book Resources → Chapter 3 → Discussion Board)*

Determining Your Desired State

Reflect on the current state of climate education and public attitudes regarding climate change. To what degree is your school, department, or community addressing climate change? Which of your answers affirm or challenge the YPCCC public opinion maps presented earlier and why do you think that is the case? What vision do you have for the future of science education? The next chapter will support your continued growth in launching storylines, but remember to plan with the end in mind. Think about your **desired state**, and what steps will be needed to get there. What will you need to accomplish with your students and colleagues as you take on the complexities of climate change? To help organize the content in this chapter, choose a curriculum map template to help create storylines that follow your school and district parameters (see Appendix B). Please note that the term "unit(s)" will be used interchangeably with "instructional segment(s)" in this chapter to begin integrating vocabulary used in the NGSS. First, access your school's science curriculum map (if one exists) for major units or topics that you are required to cover for NGSS. This might be something you have little control over, but important to seek out. You might not be able to control the flow of the units, but you can control how the segments are taught in your class. You can also center your curriculum around climate change, environmental literacy, human impact, or impact on humans to fulfill the NGSS.

Pause here to create a skeletal outline of big ideas using a curriculum map template of your choice (see Appendix B for options or select from templates provided by your school district).

The next step is to identify all the major topics for your subject area/grade level that you're required to cover. Now it's time to add new ideas and concepts to your map (see examples in Appendix C). One way to do so is to add post-it notes of essential climate change content you already know or are currently curious about, any supporting resources, and phenomena that might work well in that instructional segment. The next segment will provide key ideas of climate science, and the additional teacher resources at the end of the chapter will provide example phenomena to help launch your units. Once you feel that you have included enough, review all the post-it notes in that category to see how they might flow from one idea to another and from concept to concept. Do you notice immediately which concepts are revisited across topics? Can you identify how ideas expand each time you teach about them across your instructional segments? Lastly, take a look at all the phenomena you organized for one instructional segment and identify *one* anchoring phenomenon big enough to connect all the ideas together (directly or indirectly). See Appendix C for iterative examples provided by other teachers from the climate change educational program.

 *Before you continue to the next segment, go online to www. EmpoweredScienceTeachers.com to select a blank curriculum map to use to organize the following content, ideas, or resources.*

Climate Science Fundamentals

To understand the basics of climate science, it is essential to understand the difference between **climate** and **weather**, climate in relation to **Earth's systems**, and **how climate has changed** due to both natural forces and human activities. Climate refers to average weather conditions (precipitation, temperature, wind, etc.) with consideration for extremes that regions may experience throughout the year. Any fluctuation in temperature, rainfall, snowfall, or wind that only lasts for hours, days, or weeks is considered weather. Earth's average surface temperatures allow for an abundance of water and the thin layer of atmospheric gasses keeps our planet warm enough to sustain life. Refer to Table 3.2 to get clarification on common terms used in climate science. A common analogy is thinking about your entire wardrobe as reflecting the "climate," but your daily clothing choices reflecting the "weather" that fluctuates. You might have outfits for when weather changes, but your overall wardrobe caters to the typical climate of where you live.

TABLE 3.2 Clarification on Terms

Vocabulary	Description
Anthropocene	The geological age where human activities are the dominant influence on the environment and Earth's systems.
Anthropogenic	Resulting from human actions or activities.
Atmosphere	The layer of gasses surrounding a planet.
Biosphere	All living and nonliving organisms on Earth.
Carbon Cycle	The cycling of carbon between Earth's systems.
Climate	The average weather conditions (precipitation, temperature, wind, etc.) with consideration for extremes that regions may experience throughout the year.
Climate Change	The extra heat energy from the increased greenhouse gasses causes other changes too (such as sea-level rise, changing rain patterns, collapse of ecosystems, ocean acidification, larger than ever storms, etc.), that can have large impacts on all of Earth's systems. This term captures all those changes as a result of global warming.
Geosphere	Referring to the solid Earth.
Global Warming	An increase in global temperatures, which is caused by an increased greenhouse effect that traps more heat energy in the Earth's systems (Compare with Climate Change).
Hydrosphere	Referring to all water on Earth (oceans, lakes, rivers, streams, water vapor).
Thermohaline Circulation (Earth's Great Conveyor Belt)	Cold dense North Atlantic sea water circulates between warmer continents near the equator causing water to rise to the surface (upwelling). The water continues to travel to Antarctica where it cools, becomes more dense, and sinks again as it circulates back to the North Atlantic sea. Note that the warmer waters lose their nutrients, so this cycling supports marine life by bringing nutrients to the bottom of the ocean as colder—more dense waters.
Weather	Any fluctuation in temperature, rainfall, snowfall, or wind that only lasts for hours, days, or weeks.

Earth's Energy Budget

In my interview with a senior specialist at NASA, I wanted to know the most important concepts to teach when addressing climate change in science classrooms. The first thing that came up was teaching about Earth's Energy Budget. Understanding how energy from the Sun flows in and out of Earth's atmosphere, where the energy travels, what happens to it, and seeing how Earth balances the radiant energy are all essential to understanding climate science. Earth's systems are constantly trying to balance energy received from the Sun. When natural (such as volcanic eruptions) and anthropogenic phenomena (such as burning fossil fuels) release energy and super greenhouse gasses (GHG) into the atmosphere different chemical processes occur and at different rates. As a result, temperatures increase or decrease to reach a state of equilibrium over time. The temperatures we feel today are a result of the fossil fuels burned at enormous rates at the start of the Industrial Revolution of the 1950s. You might be wondering why that is. That's because the half-life of carbon dioxide when released into the atmosphere can be anywhere between 300 and 1,000 years (the number of years it takes to decay to half the original amount through chemical processes!), which means that the effects of climate change that we are feeling today are from emissions many years ago made by others. Imagine the exacerbated impacts for future generations anticipated given the fact that humans have emitted even more greenhouse gasses today than ever before (wait until I tell you about feedback loops!).

There are four major paths that the Sun's energy can take which may or may not result in heat radiating back to space (see Figure 3.6 provided by NASA to see the Earth's energy budget).

- ◆ Path 1: Rays enter the atmosphere and are absorbed in the atmosphere or clouds. Some of the heat radiates back to space.
- ◆ Path 2: Sunlight directly hits the ground and heats the ground surface. Some of the heat radiates back to space.
- ◆ Path 3: Rays directly reflect back to space by reflecting off the atmosphere, clouds, or the ground.
- ◆ Path 4: It's important to note that even when the Sun's rays cannot be seen during nighttime, heat continuously radiates out to space causing the environment to become colder.

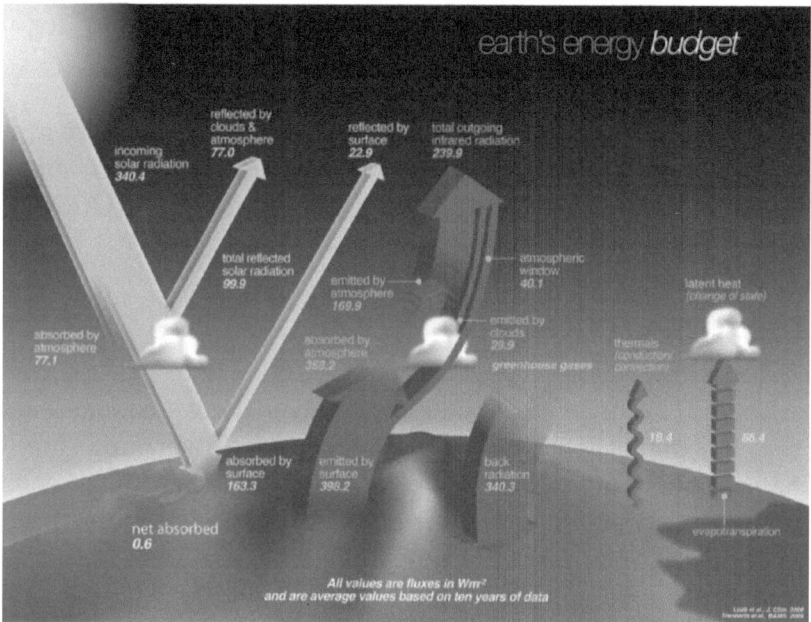

FIGURE 3.6 NASA's Visual on Earth's Energy Budget
Image Credits: NASA

Recall that learning about the Sun's pathways will allow for students to understand how some energy does not radiate back to space causing the planet to trap excess energy resulting from excess greenhouse gas (GHG) emissions.

The gasses in the atmosphere act like a heat-trapping blanket wrapped around the Earth. When we produce excess greenhouse gasses (such as carbon dioxide, nitrous oxide, methane, etc.), human activities thicken the blanket and as a result trap more heat. Imagine laying in bed under the covers and breathing under the blanket. Eventually you will increase the temperature under the blanket and it will become hot causing you to want to cool down by taking the blanket off. As we increase the heat in the atmosphere due to anthropogenic factors, Earth is trying to balance the excess hot temperatures causing a variety of changes throughout Earth's systems (such as melting ice caps, sea-level rise, ocean acidification, higher heat indexes and humidity levels, and much more). The difference is that Earth is overheating without a clear way to "remove the blanket" for all the heat trapped underneath.

 *Take it back to the class by utilizing other evidence-based communication strategies from the National Network for Ocean and Climate Change Interpretation (NNOCCI).*

Global Climate Change

Although skeptics argue that climate has always changed and that the planet is going through a period of warming due to natural phenomena such as volcanic activity, scientific research reveals however that the current rate of change is ten times faster due to human activities. If we measure Earth's average temperature beginning roughly 9,000 Before the Common Era (BCE), temperature reached modern levels and remained at roughly between 0°C and 1°C. It stays in this range and dips down to roughly −1°C and 0°C between 1CE and the 1850s at the start of the Industrial Revolution. With the introduction of fossil fuels, carbon dioxide (CO_2) emissions begin to rapidly increase in the atmosphere along with other super GHG. The start of The Great Acceleration of the 1950s began to shift Earth's average temperatures due to human activities. It is now trending toward 1°C, and our current predicted path is an increased 4°C by the year 2100 if we don't intervene. With that climate model, we are looking at the collapse of ecosystems, endangerment of marine species, sea-level rise threatening coastlines, more extreme weather events, impacts on agriculture, and much more (USGCRP, 2023). Figure 3.7 is a visual showing the timeline of Earth's average temperatures to help people understand the significantly increased rate of change over the centuries. As you explore the visual showing the drastic shift in temperatures since the Industrial Revolution, this is a great way to debunk the misconception that climate has always changed and that humans have not contributed to the current increase in greenhouse gasses.

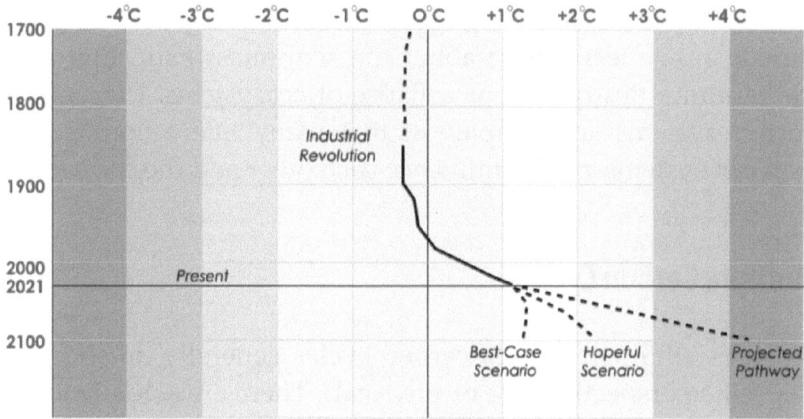

FIGURE 3.7 Earth's Average Temperatures

Earth's Systems

The Earth's climate should also be understood as a complex system where many parts interact with one another to create climate conditions. We are focusing on four primary systems that make up the Earth's systems often referred to as the **atmosphere, hydrosphere, geosphere,** and the **biosphere**. The atmosphere refers to an invisible blanket of gasses surrounding the Earth that contains gasses such as nitrogen and oxygen, and smaller amounts of trace gasses such as water, carbon dioxide, methane, and ozone. All of these gasses influence Earth's climate system regardless of their small quantities. The hydrosphere refers to the water on Earth's surface (in liquid, solid, or gas form). Water stores a great deal of heat and absorbs large amounts of CO_2. The oceans circulate the heat around the globe through ocean currents, which transfers energy to the atmosphere playing a large role in influencing Earth's climate. The geosphere refers to Earth's land from the surface down to the core. Rocks play a large role in shaping Earth's climate because of how much carbon they can store, and volcanic eruptions release a great deal of particles and gas into the air impacting Earth's climate. Lastly, the biosphere refers to all living things on Earth. Living things impact each system because of the amount of carbon they can emit or absorb. Living things are also impacted by the changing climate as we learn more about the sixth-mass extinction and the resulting destruction or collapse of ecosystems. Ultimately, Earth's systems are complex in that many interactions occur between systems as they influence each other and the climate.

Multiple Carbon Cycles

Teachers who teach about carbon cycles generally do so from a specific perspective (life or physical). There are actually multiple perspectives of the carbon cycle that occur because carbon moves through each of the Earth's systems in different ways. When individuals fully understand the impacts of their daily decisions, they begin to care about and take action on the problem. The carbon cycles can help students understand the source and impacts of excess anthropogenic emissions that are causing ocean acidification, dissolving of the sea floor, damaging key marine species like phytoplankton who provide nearly half the oxygen for Earth, melting of the permafrost, among many other harmful impacts. This segment will provide an overview of the various carbon cycles to help you see the larger picture.

The Carbon-14 Dating Cycle (A Physical Perspective)

The Carbon-14 (C-14) dating cycle allows for students to understand how carbon enters living things and how it could be released due to anthropogenic activities (such as burning fossil fuels or drilling for natural gas). Scientists can use C-14 radioactive dating to figure out how old something is based on when it died. They do so by analyzing the element's half-life in remaining bones, wood, or cloth samples. This process starts with the Sun's rays colliding with atoms in the atmosphere to create high energy neutrons and eventually moving to living species consuming the carbon-rich nutrients. Toward the end, the cycle is complete when decomposers work to put nitrogen back into the atmosphere from decayed matter (see Figure 3.8 for more details). As living things decay and are buried over time, they

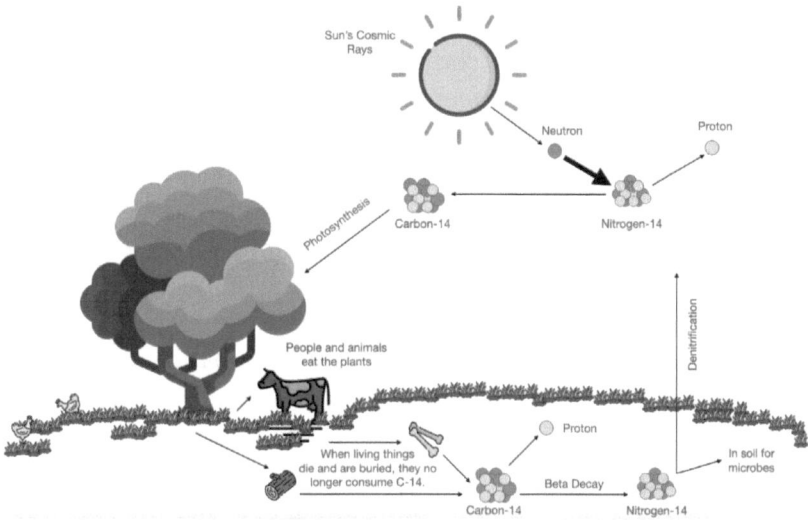

FIGURE 3.8 Carbon-14 Radioactive Dating Cycle

store carbon that remains virtually undisturbed (with exceptions of both natural phenomena and human activities).

The Marine Carbon Cycle

The marine carbon cycle has three paths that interact with each other allowing some carbon to return to the atmosphere. This cycle helps students to understand how the ocean absorbs at least a quarter of the carbon released (in the form of CO_2), and acts as a circulatory system for Earth. The ocean is also known as Earth's "heart" because it helps the planet to reach equilibrium by absorbing the excess heat in the atmosphere. As the ocean cycles heat and nutrients across the planet through thermohaline circulation (the great ocean conveyor belt), it works as a regulatory system for Earth.

Path 1: Carbon dioxide from the atmosphere dissolves into the ocean and sinks to lower depths with colder waters. Through thermohaline circulation, the CO_2 eventually rises up as bubbles with warmer waters due to crashing ocean waves and returns to the atmosphere. This process is slow and relies on the turning of the ocean waters as it travels from the equator toward the poles (View NASA's animation at www.Bit.ly/NASATC). Scientists have been warning us that the conveyor belt is beginning to slow and weaken due to climate change. These are impacts we are currently seeing and feeling through colder winters, hotter summers, direct impact on water and food supplies, and more.

Path 2: Carbon dioxide from the atmosphere dissolves into the ocean, and is absorbed into phytoplankton and algae as they engage in the process of photosynthesis (similar to how plants do this on land). Through the food chain, some of the carbon returns to the atmosphere when sea creatures participate in respiration. When sea creatures die in the ocean and sink to the bottom of the ocean they decay. The bacteria begin to decompose the dead sea creatures and breathe out CO_2 that returns carbon to the atmosphere in the process.

Fossil fuels (like oil and natural gas) formed in the ocean are the result of dead marine species (such as plankton) that have been buried by mud and sand. The remains change into oil due to high pressure, high temperatures, and the help of bacteria. The remains can also change into limestone, which is commonly extracted to make concrete. These non-renewable sources of energy take millions of years to develop under specific conditions. The carbon would normally be stored underground, but humans have been extracting fossil fuels for energy. That means that when we burn fossil fuels (which is anthropogenic), we are releasing excess amounts of carbon into the air that otherwise would have been stored underground.

Path 3: This path explains ocean acidification and how the ocean absorbs excess CO_2. The carbon dioxide in the atmosphere dissolves in the ocean producing carbonic acid (H_2CO_3). Due to the nature of acids, carbonic acids can lose a hydrogen-ion when reacting with water which converts it into bicarbonate-ion (HCO_3^-). When bicarbonate-ions (which are acidic) lose a hydrogen-ion it turns into carbonate-ions (CO_3^{2-}). Marine species (such as lobsters, crabs, phytoplankton, etc.) rely on carbonate-ions to create their shells.

There are equilibrium reactions that occur between these three chemical reactions as well (see Figure 3.9). If there are lower levels of CO_2 in the atmosphere compared to the level in the ocean, carbonic acids will release carbon back into the atmosphere and

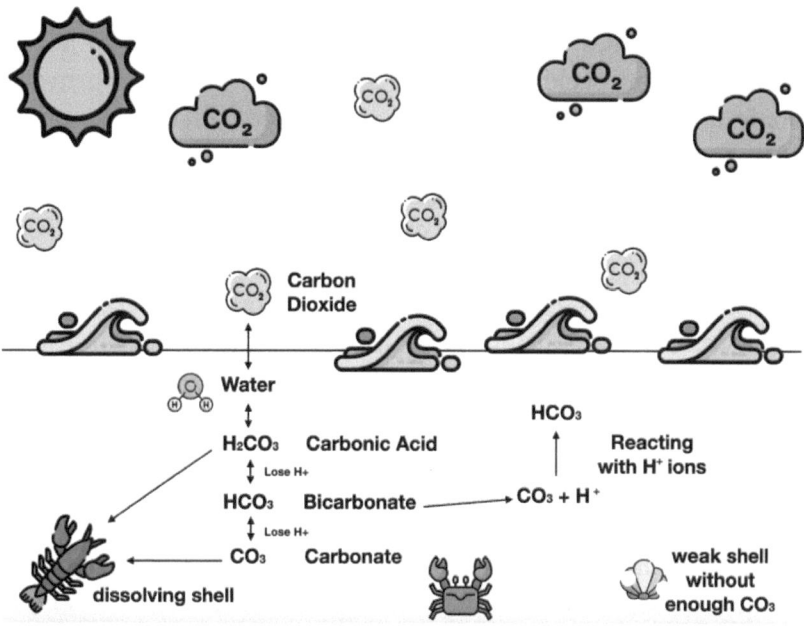

FIGURE 3.9 Carbon Cycle & Ocean Acidification

the chain reaction described previously occurs. On the other hand, if there are increased levels of CO_2 in the atmosphere then more carbon is dissolved into the ocean forming more carbonic acid, bicarbonate-ions, and carbonate-ions. The carbonate-ions in the water continue to react with dissolved CO_2 to form more bicarbonate-ions which allows the ocean to hold 10 times more carbon dioxide (known as the Ocean Buffer). Remember that carbonate-ions are needed for marine species, but when carbonate reacts with CO_2 it produces bicarbonate-ions (which cannot be used for shell production). As a result, shells formed are weaker without enough carbonate-ions, and the current shells begin to dissolve because of the carbonic acid in the ocean. This is known as ocean acidification (see NOAA for more www. Bit.ly/NOAAoceanacid).

The Geologic Carbon Cycle

Another carbon cycle to consider occurs in the geosphere, where volcanic eruptions or metamorphic events release CO_2 into the atmosphere naturally. Rocks in the ground play a big role in storing carbon and other gasses through the rock cycle (Access NASA's resource at www.Bit.ly/NASAROCK). Similar to the marine carbon cycle where CO_2 reacts with water to form carbonic acid, this happens when rain water comes into contact with rock sediment starting the chemical weathering process. As rivers carry the various ions into different bodies of water (lakes, streams, oceans, etc.) a chemical reaction occurs. The sediment builds up over time and buries any dead marine specimen over time. Layers continue to build through this process, and the increased pressure turns shells and sediment into limestone allowing for carbon capture.

Feedbacks, Sources, and Sinks Oh My!

If we want to focus on solutions, it's important to understand **feedback** loops that reinforce or stabilize our current system, as well as **sources** of emissions and major **carbon sinks**. In general, "positive feedbacks" are actually bad because they reinforce the process that can lead to even more warming. An example of this can be seen with snowmelt. When the temperature warms up, it melts the white snow which is completely expected. As it continues to melt, however, it begins to reveal the darker colored surface or ground underneath which doesn't reflect the sunlight as well as the white snow and begins to absorb the Sun's heat (ChatGPT the albedo effect and permafrost). With the ground being uncovered and warmer, it begins to melt even more snow and faster. It's a reinforcing loop because the darker surfaces that absorb more heat leads to even more melting and warming. A stabilizing or "negative" feedback loop can be seen through our one and only ocean. The ocean absorbs massive amounts of CO2 from the atmosphere and as the gasses dissolve at the surface of the waters, it undergoes chemical reactions that impact the acidity level of the ocean as well as the marine species that inhabit it. The carbon is eventually stored as carbonate sediments for potentially thousands to millions of years (depending on geological processes). The ocean helps to mitigate the climate crisis and acts as a **carbon sink** (effectively storing, absorbing, or removing

EMISSIONS SOURCES & NATURAL SINKS

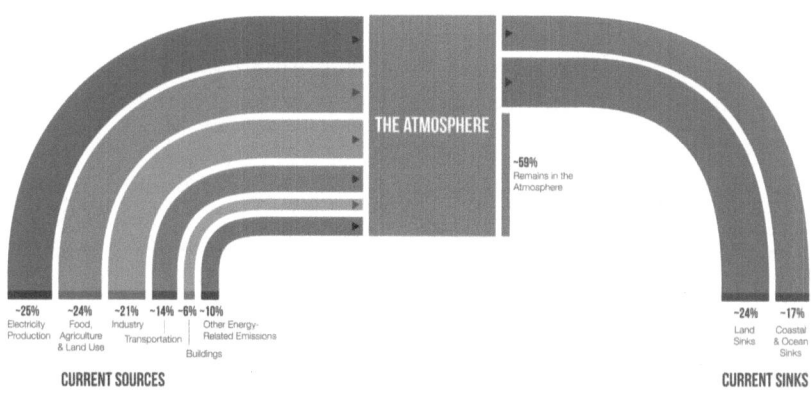

FIGURE 3.10 Carbon Sources and Sinks
Source: Project Drawdown, drawdown.org.

of carbon to help regulate and stabilize global temperatures for long periods of time). It's important that we understand, however, that the ocean will reach critical tipping points in which the ecosystems thriving in it will begin to degrade as a result of ocean acidification and other impacts if we don't change our thinking and behaviors.

Project Drawdown emphases the need to identify and reduce large sources of carbon emissions in order to stop them, supporting current sinks (e.g. advocacy for mangroves, seagrass, coral reefs, the ocean, ancient forests, etc.), and improving equality for all living things. Review Figure 3.10 to see major societal sources and sinks.

The Anthropocene

Now that you understand both natural and anthropogenic phenomena taking place through the energy budget and various carbon cycles, how do we know what the causes of current climate change are? It's humans. We are such a force of nature as a species that we are able to alter every Earth system with our actions (and inaction). Recall that this geological time period marked by human actions is known as The Anthropocene. "The Age of Humans" is a result of the Industrial Revolution, Great Acceleration, and dramatic population growth (Ramanathan, Aines, Auffhammer, Barth, Cole, Forman, et al., 2019). These advancements have allowed humans to thrive and make technological advances for a better life on Earth, but we have done so at the expense of everything else (Table 3.3).

TABLE 3.3 Teaching about the Anthropocene

Topic	Description
Climate Reality Project	Check out this resource for ways to teach students about anthropogenic climate change.
	Link: www.Bit.ly/CRPHUMANS
UC-CSU NXTerra— Learning Resources	Review a variety of resources that provide education, tools, and teaching tips on approaching the climate crisis.
	Link: www.Bit.ly/TEACHINGGCC
NASA—Images of Change	Have students analyze images from across the globe on the impacts of climate change.
	Link: www.Bit.ly/PicChange
Our Climate Our Future	Explore teaching resources and student videos to introduce the causes of climate change and ways to get involved through the Alliance for Climate Education.
	Link: https://ourclimateourfuture.org/
Specific Topics to Explore for Concrete Examples of Anthropogenic Factors	Climate change linked to consumerism.
	Production and the role of plastics (such as plastic bottles, bags, take out containers, etc.).
	Deforestation, reliance on palm oil, and rise of diseases.
	Burning Fossil Fuels.
	Extraction and uses of limestone.
	Travel (commuting for work or for leisure)
	Food production and transportation.
	The raising and consumption of cows, pigs, chickens, etc.
	Deep sea ocean or permafrost drilling.
	The role and effects of fracking.

Current climate change is anthropogenic because the excess GHG emissions are a direct result of human activities and actions. One way to help students understand anthropogenic climate change is to have them compare regular and rampant CO_2 emissions. The NGSS calls for teachers across science subjects and grade levels to teach the carbon cycle pathways. This is one way to showcase natural events that release carbon dioxide into the atmosphere compared to the excess levels of GHG that humans emit due to daily activities. It may be shocking for students to learn that humans emit almost 100 times more carbon dioxide than all volcano eruptions combined, so the recent increase in temperatures as a result of excess CO_2 cannot be attributed to natural phenomena (Ramanathan et al., 2019). If people do not learn climate science and fail to see how scientific data supports the claim that humans are the cause, they will not feel the urgency of this crisis or take the necessary actions needed to bend the curve. See Table 3.3 for additional resources on how to teach anthropogenic climate change.

Scientific Consensus on Climate Change

Although the NGSS requires secondary science educators to teach climate change content, teachers report avoiding the topic because they considered it highly controversial and heavily debated among the scientific community (Hestness et al., 2011; Liftig, 2012; Maibach et al., 2014). When science educators receive climate change information from media sources, the information may not reflect scientific research or may even portray climate science as controversial. Maibach et al. (2014) found that 42% of Americans believe that most scientists think that climate change is happening, and 33% believe that there is no scientific consensus among climate scientists. This is problematic when 97% of the scientific community agrees that climate change is occurring due to human related activities (Cook, Nuccitelli, Green, Richardson, Winkler, Painting,... & Skuce, 2013, Cook, Oreskes, Doran, Anderegg, Verheggen, Maibach et al., 2014).

There's actually **no** debate about whether climate change is happening among scientists. Instead, a small number of researchers (less than 3% and many in fields outside of climate science) are questioning whether it is due to anthropogenic causes because the data collection is still ongoing. Teachers should not have students debate about climate change because it sends the message that personal opinions outweigh credible data gathered by climate experts. Debates should instead be about policy or ethical dimensions of climate change (such as the growing number of climate refugees, the growing food crisis, what to do about the collapse of ecosystems, sources of energy for the future, etc.), but not about the science of climate change.

 *Take it back to the class with the following resources to introduce scientific consensus to students to debunk misconceptions circulating through the media.*

1. *Go to www.Bit.ly/CookConsensus for teacher resources provided by Skeptical Science.*
2. *Show students John Oliver's Mathematically Representative Climate Change Debate—Bit.ly/CCDEBATE*
3. *Have students analyze Cook's article for data on consensus— www.Bit.ly/COOKconsensus*

Major Teaching Takeaways

Although the following five points are layered with a great deal of content, these are the major teaching takeaways for any science teacher looking to teach climate change.

1. **Teach the scientific consensus on climate change**—This is the perfect opportunity to teach elements of the Nature of Science, argumentation in science using credible evidence to support a claim, and provide clarification for students on how this topic is politically controversial but not scientifically.

2. **Do not debate about climate change**—Allowing students to debate about climate change sends the message that their opinion trumps credible scientific evidence and data in the end. Climate change debates should focus on the ethical or political dimensions of the climate crisis after learning about scientific consensus (For example, should we continue to rely on palm oil for our lower-priced processed foods or try to find another solution that doesn't rely on deforestation, is more sustainable, and causes less harm?).

3. **Teach climate change through the lens of systems thinking**—Consistently come back to the crisis throughout different segments and lessons to show how interconnected and interdisciplinary science and the impacts of climate change are (ex. Direct impacts on the hydrosphere also impact the other spheres which have rippling effects or how everyday plastics end up in the ocean and eventually back into your body through different sources). Remember that concepts or ideas not revisited often will not be retained or transferred by students. The NGSS also stresses teaching science through systems-thinking to show the cyclical and iterative Nature of Science.

4. **Teach climate change as a socioscientific issue**—Research reveals that this approach to climate science is effective in helping students to develop scientific literacy skills by learning more about the Nature of Science. To do so, (1) Consider how you design your lessons around climate change phenomena, (2) be open and willing to learn and teach about a complex subject, and (3) integrate opportunities to address the ethical and social dimensions related to the topic because it directly impacts students and has the ability to activate students as agents of change (For example, students could analyze data and pose higher-level questions interacting vetted online tools showing direct community impacts).

5. **Teach the cause of current climate change with the impacts**—The NGSS emphasizes the need to integrate "human impact" through all sciences in iterative ways. As teachers take on education for climate action, they first need to teach about anthropogenic causes. Put simply, if humans are the cause then we're also the answer. We can all be part of the solution in different ways that need to reflect our diverse nature. Students will want to know what they can do to help adapt or mitigate climate change, and understanding that their direct actions are tied to the climate crisis might influence their daily decisions.

6. **Knowing climate science and teaching about it are two very different skills.** How you approach climate science is just as essential as what you decide to include. If data and information alone was enough to bend the curve, we wouldn't have a climate emergency on our hands. Leaning into culturally relevant and responsive teaching, applying the teaching components of climate as a socioscientific issue, providing opportunities to unveil and see potential solutions to addressing the challenges and injustices are huge learning moments, and humanizing the crisis to tend to students' social and emotional needs helps to bridge the gap between knowing and taking action.

As you consider what science content to include or exclude in your existing or newly developing curricula, think about how you can anchor storylines around climate and environmental topics. This will allow you to build on students' interests, their lived experiences, funds of knowledge, critical thinking skills, and so much more. Recall from previous chapters that incorporating the content is one way to address the NGSS framework, but the complementing piece is thinking through the delivery of that lesson to build students' capacity as future leaders and problem solvers that can apply what they have learned. The next chapter will go through pedagogical practices and resources to successfully anchor your curriculum on climate change.

Collective Voices for Climate Change Education

Jennifer Cao, Special Projects Coordinator at UC-CSU ECCLPs

Humans have a collective responsibility to protect and preserve the environment for current and future generations. As teachers, we are in a special position of power and privilege to provide engaging, informative, and inspiring learning experiences. Climate change is the perfect platform where we can support students' critical thinking skills while addressing environmental injustices and social inequities. Leveraging climate change provides an opportunity for students to embrace the shared responsibility of stewardship of our planet!

Richard Vevers, Founder and CEO of The Ocean Agency

Children naturally encounter information about climate change, but the narratives they usually come across can provoke fear and anxiety during their formative years. It's crucial they learn about this issue in a safe and supportive environment. The knowledge can change how they look at the problem and empower them with a greater sense of control and hope. Moreover, discussing solutions and progress not only promotes resilience but also fosters optimism. That's why teaching about climate change is so important for their wellbeing.

Additional Teacher Resources

A year in the lift of Earth's CO_2 emissions video—
  Bit.ly/GHGNASA
Access "Bending the Curve" to learn more about climate
  solutions—
  Bit.ly/BTCBOOK
Access TedEd United Nations online environmental lessons—
  ed.ted.com/earth-school
Curious about how your state is "Making the Grade" on climate
  education?—
  climategrades.org/
Check out HHMI's Geologic Carbon Cycle resources—
  Bit.ly/HHMICarbon
Explore the NxTerra website for climate change resources—
  Bit.ly/NXTerra
Explore the California UC-CSU ECCLPs initiative—
  www.ecclps.net
Get free project-based learning curricula on environmental
  topics from SEI—
  Bit.ly/TEACHSEI
Learn more about the Anthropocene here—
  www.anthropocene.info/
Look over the NCSE website for teacher resources—
  ncse.ngo/teaching-climate-change
Review this study showing the flaws of climate skeptics'
  research—
  Bit.ly/StudyFlaws
Subject to Climate has a great deal of resources to offer—
  subjecttoclimate.org
Teach using the full visual of Earth's average temperatures by
  XKCD—
  Bit.ly/XKCD1732
Use the CalAdapt tool to see how your community will be
  impacted—
  cal-adapt.org/
Use this tool to see the real-world cost of climate change—
  http://www.impactlab.org/

View national Yale Climate Opinion Maps—
  Bit.ly/YaleMaps
Watch "Before the Flood" documentary—
  www.beforetheflood.com/
Watch "Chasing Coral" documentary—
  www.chasingcoral.com/

References

Cook, J., Nuccitelli, D., Green, S. A., Richardson, M., Winkler, B., Painting, R., ... & Skuce, A. (2013). Quantifying the consensus on anthropogenic global warming in the scientific literature. *Environmental Research Letters*, 8(2), 024024.

Cook, J., Oreskes, N., Doran, P. T., Anderegg, W. R., Verheggen, B., Maibach, E. W., ... & Nuccitelli, D. (2016). Consensus on consensus: A synthesis of consensus estimates on human-caused global warming. *Environmental Research Letters*, 11(4), 048002.

Crimmins, A.R., Avery, C.W., Easterling, D.R., Kunkel, K.E., Stewart, B.C., & Maycock, T.K. (Eds.) *USGCRP, 2023: Fifth National Climate Assessment*. Washington, DC: U.S. Global Change Research Program. https://nca2023.globalchange.gov/credits/

Frame, A. (2020, April 21). *Climate Change: It's Not One More Thing-It's the Thing*. Retrieved October 15, 2020, from https://tenstrands.org/ci/climate-change-its-not-one-more-thing-its-the-thing/

Goldberg, M., Gustafson, A., Rosenthal, S., Kotcher, J., Maibach, E., & Leiserowitz, A. (2020). *For the first time, the Alarmed are now the largest of Global Warming's Six Americas*. New Haven, CT: Yale University and George Mason University; Yale Program on Climate Change Communication.

Hestness, E., McDonald, R. C., Breslyn, W., McGinnis, J. R., & Mouza, C. (2014). Science teacher professional development in climate change education informed by the next generation science standards. *Journal of Geoscience Education*, 62(3), 319–329.

Hestness, E., Randy McGinnis, J., Riedinger, K., & Marbach-Ad, G. (2011). A study of teacher candidates' experiences investigating global climate change within an elementary science methods course. *Journal of Science Teacher Education*, 22(4), 351–369.

Howe, P., Mildenberger, M., Marlon, J., & Leiserowitz, A. (2015). Geographic variation in opinions on climate change at state and local scales in the USA. *Nature Climate Change*. https://doi.org/10.1038/nclimate2583.

Lambert, J. L., & Bleicher, R. E. (2013). Climate change in the preservice teacher's mind. *Journal of Science Teacher Education*, 24(6), 999–1022.

Liftig, I. (2012). A tough climate for teachers. *Science Scope*, 35(7), 1–1.

Liu, S., Roehrig, G., Bhattacharya, D., & Varma, K. (2015). In-service teachers' attitudes, knowledge and classroom teaching of global climate change. *Science Educator*, 24(1), 12.

Maibach, E. W., Leiserowitz, A., Roser-Renouf, C., & Mertz, C. K. (2011). Identifying like-minded audiences for global warming public engagement campaigns: An audience segmentation analysis and tool development. *PLoS One*, 6(3), e17571.

Maibach, E., Myers, T., & Leiserowitz, A. (2014). Climate scientists need to set the record straight: There is a scientific consensus that human-caused climate change is happening. *Earth's Future*, 2(5), 295–298.

Plutzer, E., McCaffrey, M., Hannah, A. L., Rosenau, J., Berbeco, M., & Reid, A. H. (2016). Climate confusion among US teachers. *Science*, 351(6274), 664–665.

Project Drawdown, 2021. *Project Drawdown*. (2024, February 29). https://drawdown.org/

Ramanathan, V., Aines, R., Auffhammer, M., Barth, M., Cole, J., Forman, F., et al. (2019). *Bending the Curve: Climate Change Solutions*. Regents of the University of California. Retrieved from https://escholarship.org/uc/item/6kr8p5rq

Sadler, T. D., Chambers, F. W., & Zeidler, D. L. (2004). Student conceptualizations of the nature of science in response to a socioscientific issue. *International Journal of Science Education*, 26(4), 387–409.

Climate Change as the Anchor

Read this when:

- ◆ *You understand the fundamentals of climate science and are ready to integrate more content into your curriculum.*
- ◆ *You want to learn more about how to build students' scientific literacy and socioscientific reasoning skills.*
- ◆ *You are ready to learn about phenomena-based instruction to continue creating meaningful storylines.*

Lessons from Bonsai Koi Fish

Did you know that bonsai koi fish that live in healthy natural ponds can grow up to two feet in length, but those that grow in man-made tanks will not outgrow the container they live in? The most notable effect is that confined koi fish experience stunted growth due to environmental factors that would otherwise allow them to grow much larger. As a result, they also have shorter life spans than those that live in healthy natural ponds with thriving ecosystems. Pause here to think about our educational system. Remember that education is a cultural act. It's not universal and it's created by people who determine what that environment will or won't have. So who gets to determine what and how much students do or don't learn? Remember that education is a cultural

DOI: 10.4324/9781003478584-7

act. It's not universal in approach and it can be greatly influenced by those who make up the system. Let's reflect together and think about whether we are willing to leverage our decision-making power in the classroom to disrupt traditional science education so that students learn 21st-century skills for the problems they face in their communities *now*. Can we revamp this system to support every learner as NGSS and our society demands to combat urgent issues such as the climate crisis?

It's no secret that students rely heavily on their teachers as trusted sources for data and information. Don't forget that teachers hold a great deal of power in the classroom as individuals who determine what will be taught, how it will be taught, how students will be positioned, what is deemed worthy of time, and so much more. Given this important role, it is essential to support teachers with the paradigm and pedagogical shifts needed to teach socioscientific issues such as climate change.

There is a vast amount of research that reveals how highly dependent people are on the media and internet to learn about global climate change (Caranto & Pitpitunge, 2015; Carter & Wiles, 2014; Hansen, 2010; Hestness, McDonald, Breslyn, McGinnis, & Mouza, 2014; Hodson, 2003; Matkins & Bell, 2007; Somerville & Hassol, 2011). Among these individuals are teachers and students that also rely heavily on social media to get scientific information. Remember that climate change is an SSI because students are bombarded with messages and claims about it, whether teachers actively address it or not. **What's clear is that when teachers choose to avoid or omit climate change altogether, they are indirectly sending messages to students about their underlying core values and beliefs** (Kolstø, 2001; Sadler, Chambers, & Zeidler, 2004).

Just as antiracist science teaching requires an active and intentional effort rather than choosing the safety of silence (Kendi, 2019), teaching about climate science requires a deliberate effort. This is an opportunity to engage students in learning about the science of climate change to provide them with a safe space to ask questions, debunk misinformation, express their emotions, model lifelong learning, use data and credible evidence to support their claims, engage in science and engineering practices, activate their agency to take action on direct impacts to their communities, and so much more. Teachers are powerful agents of change, and these are great opportunities for students to experience uninhibited growth.

100Kin10 Predictions

100Kin10 is a national organization that seeks to provide students with high-quality STEM education through recruiting and retaining 100,000 excellent STEM teachers by 2021. Through their work and data collection, they have identified seven grand challenges underlying the STEM teacher shortage that require solutions. These include professional learning, teacher preparation, ensuring schools value the S, T, and E in STEM, and several others. In their 2019 Predictions Report, 100kin10 predicts that environmental advocacy will engage more students in STEM. Worldwide, students are taking part in movements and opportunities for activism that address environmental challenges such as climate change. Students are taking the lead and the data reveals that schools need to start responding to these growing interests to support these efforts.

The NGSS has the ability to catalyze climate change education, while enhancing scientific and environmental literacy efforts as an intersectional issue. This chapter will provide you with the tools, knowledge, and resources needed to tap into your students' agency. As the desire to mobilize on climate change increases among our nation's youth, teachers can support that growth and desire with purposeful curriculum design.

Asking More Questions

There's a really interesting biological perspective on the need for diversity in thinking around climate resilience and solutions. In nature, ecosystems thrive with increasing biodiversity and they can completely collapse or destabilize due to the lack of diversity. The interactions among and between the species plays a critical role in the features, functions, and resilience of the ecosystem. Whether the decline is a direct result of habitat destruction, pollution, invasive species, or other factors, biodiversity is essential for a thriving biome. Take this into context with limited diversity of thinking and solutions for the climate emergency. Without diversity, we cannot respond effectively to the complex challenges to our entire system. Even worse, the narrow perspectives from individuals that have the power to make decisions for all might create even more injustice and inequity as a result of the lack of diverse pathways that can identify multiple solutions and opportunities to mitigate or adapt to climate change. Solutions are only solutions so long as they are used, otherwise they are all but theoretical (especially if they disregard context, culture, and communities).

In Chapter 3 you learned that climate change is a complex topic without straightforward solutions welcoming diversity of thinking and teaching. As a result, your students will ask bigger and more complex questions that you may or may not have the answers to at the moment. Anticipate their questions and where they may face challenges with the content, and embrace those opportunities. That's when you know they're really engaged! Recall that socioscientific issues embody elements of the Nature of Science because the data and information collected is growing and unveiled over time. You are not expected to know all there is about climate change because it has a tentative nature. With new technology and scientific advancements, this allows for the collection of new data, trends, and patterns that will need to be studied. Embrace these moments as ones where you can model lifelong learning with ways to unpack NOS principles. There is no doubt that learning more about how destructive human actions can be, will push our societies to discover more innovative ways to solve the problem both at the individual and systemic levels. Let's remain hopeful that future generations will one day read about how we were able to tackle this problem by recognizing our collective power as educators.

For your students, this is an opportunity to model what lifelong learning looks like, engage with the Nature of Science principles to position them authentically as scientists or engineers, and provide them with the skills needed to explain what is happening in the world. You might not have the answers to their questions today, but you could be witnessing a future activist take interest in a problem they want to find a solution for in the near future. This chapter will guide you through facilitating and supporting your students in developing scientific literacy skills, critical thinking capacity, and thinking through how to create storylines using climate change as the anchor.

Using SSI to Develop Scientific Literacy Skills

What Is Scientific Literacy Anyway?

Scientific literacy has many different definitions that have evolved over time. The National Research Council (NRC) defines scientific literacy as, "[T]he knowledge and understanding of scientific concepts and processes required for personal decision-making, participation in civic and cultural affairs, and economic productivity. It also includes specific types of abilities (1996)." The NRC goes on to explain that a scientifically literate person is able to ask, find, or determine answers to questions that arise from everyday experiences. These individuals are able to describe, explain, and make predictions for everyday phenomena. They are also able to evaluate the quality of scientific information and make claims based on credible sources of evidence. Lastly, these individuals are able to express positions on scientific and technological issues underlying national and local decisions as informed-citizens. It is important to note that determining whether educators have been teaching students to be scientifically literate depends on their own understanding and definition. Anelli (2011) synthesizes different definitions of scientific literacy reflected in Table 4.1. Read through the definition and determine the one closest to your teaching disposition.

 *Which definition of scientific literacy aligns most to your beliefs about teaching and learning? Why? How can you utilize climate change to develop students' scientific literacy skills according to the definition(s) you identified most with?*

TABLE 4.1 Defining Scientific Literacy

Definition by Date	Description
Practical, Cultural, and Civic Scientific Literacy proposed by Shen (1975)	"Practical" refers to the application of scientific principles or technology needed to improve life. "Cultural" refers to the appreciation of science as a human achievement. "Civic" refers to the level of understanding needed to engage in science-related issues.
Civic Scientific Literacy proposed by Miller (1983, 1998)	Expounding on the previous definition, Miller defines civic scientific literacy as the level of understanding needed to read and comprehend science and technology related issues. They should be able to engage in societal debates that involve science and technology as informed-citizens.
Functional Scientific Literacy	People need to have a foundational understanding of science because they cannot think critically about nothing. Even if most people will never hold careers in science, they will need to function as citizens and need to be scientifically literate to make informed decisions.
Scientific Literacy as Defined by NAS for the National Science Education Standards (1996)	"Scientific literacy is the knowledge and understanding of scientific concepts and processes required for personal decision-making, participation in civic and cultural affairs, and economic productivity. It also includes specific types of abilities."
Fundamental and Derived Scientific Literacy proposed by Norris and Phillips (2003)	Fundamental scientific literacy has been simplified to only include reading, writing, accessing, and synthesizing information. These scholars argue that it should also include the ability to interpret, infer, analyze, critique, and contextualize information related to science. Students need to practice analyzing scientific texts with different intentions (an observation, causal relationship, generalizations, hypotheses, assumptions, supporting evidence, etc.) so they can also apply these skills to analyzing science in the media.
Useful Scientific Literacy proposed by Feinstein (2011)	Science education needs to focus on the usefulness of scientific literacy. This includes the skills needed to solve personally meaningful, everyday problems, while also being able to make informed decisions on science-related issues. Feinstein argues that educators should teach students how to recognize science in "real-life" contexts where they can apply the scientific literacy skills.

Definition by Date	Description
Scientific Literacy in the NGSS (2014)	"One fundamental goal for K-12 science education is a scientifically literate person who can understand the nature of scientific knowledge. Indeed, the only consistent characteristic of scientific knowledge across the disciplines is that scientific knowledge itself is open to revision in light of new evidence."

Scientific Literacy in the NGSS

Recall that to support scientific literacy, the NGSS embedded NOS principles within bundled performance expectations to support students engaging in three-dimensional learning. As we build students' scientific literacy skills and capacity, this requires teachers to be able to facilitate experiences that transition students from dependent to independent learners (with scaffolds that scale back over time). Research data affirms that when students are able to apply the NOS to complex socioscientific issues (such as the climate crisis), they increase their critical thinking capacity and scientific literacy skills through deep and relevant learning opportunities (Carter & Wiles, 2014; Khishfe & Lederman, 2006; Kolstø, 2001; Matkins & Bell, 2007; Sadler, Barab, & Scott, 2007). Consider engaging with Exhibit 4.1 to reconsider or affirm some of your own teaching practices that build critical thinking capacity.

Exhibit 4.1 Reframing to Build Students' Critical Thinking Capacity

The following lessons were created with Karajean Hyde (Director of the Irvine Math Project) for the CA Partnership for Math and Science Education grant. Together we sought to develop interdisciplinary lessons that centered on climate change investigations coupled with culturally relevant and responsive pedagogy. We developed several multi-day lessons together using Desmos (www.desmos.com), which is a free interactive tool that helps students to model, build, and engage with data and graphs. To access any of the following free teaching lesson plans coupled with video support, go to www.bit.ly/MathScienceDesmos.

In just highlighting what we learned about science teacher program participants, we realized quickly that engaging, building confidence in, and teaching of rigorous math modeling in science needed more attention. Not only were teachers using a new teaching tool (Desmos), they were also learning about how to ask key questions to elicit student responses that would unveil their funds of knowledge and any potential gaps in understanding. Climate change and environmental topics were the vehicle used to support students with sense-making and specifically with math modeling. Below are the lessons we developed and a summary of how we complicated each resource to help students engage in NGSS SEPs (such as asking questions, argumentation from evidence, engaging in math and computational thinking, developing and using models, etc.).

Lesson 1 (For grades 6–8): Air Quality Factors with a Focus on Ratio and Proportions

◆ Day 1: We started this lesson by employing free tools to help students generate the anchoring question to drive the experience. We asked the students to look up air quality in their neighborhood and knew that it would result in varying numbers. We also asked students if they felt the air quality was the result of any recent wildfires when they looked over data that revealed the last 30 days of air quality from https://fire.airnow.gov/ to help them make claims based on evidence. The question we posed for them at the end of this lesson pushes them to think about whether or not the current air quality numbers are what's causing variance because another data mapping tool from 2021 showed air quality prior to any wildfires to complicate their current claims.

◆ Day 2: Students play with the free CalEnviroScreen 4.0 tool (www.bit.ly/CalEnviro) to share what they notice and wonder about the information. They move to learning more about air pollution chemistry and how it's measured to understand the data. Social justice and climate change are intersectional. Following this lesson, we encouraged teachers to provide space for students to discuss their thoughts on the fact that different communities breathe in different qualities of air and what can/should be done.

Lesson 2 (For grades 9–12): Carrying Capacity & Human Influences with a Focus on Functions

◆ Day 1: Students will explore the investigative question, "What is happening to the populations of tigers, rhinos, elephants, and orangutans in Sumatra (one of the last places on Earth where all four animals live together in one ecosystem)?" They will review data and information to learn more about the food chain, and then be encouraged to make initial claims for why the population is increasing or decreasing for each animal. Students will then be introduced to concepts of "carrying capacity" and "keystone species." It's expected that students will assume that the keystone species (the Sumantran Tiger) has been over consuming the other species, until they see the decline in population of the tiger as well to generate more wonderings. As a result, they will explore graphs of the decreasing tiger population as well as deforestation rates of the area and notice correlations. We end the lesson by asking students to share questions about what they think is happening when the keystone species is also declining in population, and if there are connections to the deforestation efforts and if so why.

◆ Day 2: Students will analyze additional graphs that show how many Hectares of forest are decimated in this area each year and CO_2 emissions from deforestation along with local community and human impacts. Students will receive an introduction to "reforestation" as one method for counteracting the adverse impacts of deforestation. Utilizing provided data on Desmos, they will contribute to the graph depicted, and anticipate the point at which reforestation will offset the level of deforestation. They will learn more about why deforestation is happening in Indonesia leading them to learn more about palm oil and its uses. Students will then think about local actions they can take to change behaviors and consumption patterns informed by what they learned in the lesson.

The collaboration with my math colleague allowed for me to learn more about how to effectively model math in science through free and engaging ways. We know that climate change is an intersectional and complex problem requiring diverse responses. Think through ways that you can connect across disciplines to provide deeper and more engaging opportunities that help move us out of our silos.

Using Climate Change to Develop Socioscientific Reasoning

As students begin to take more active roles in environmental justice issues that directly impact their lives, it is essential to develop their capacity to successfully lead changes in their communities. Rather than teaching climate change as isolated topics throughout the storyline, climate science can be used to ground the class as the larger issue students create solutions for. Approaching climate change through the SSI framework allows for students to engage in *socioscientific reasoning* that increases their content knowledge and understanding of the NOS for scientific literacy (Sadler et al., 2007). **Socioscientific reasoning** includes being able to recognize the complex nature of SSI, examine issues from different perspectives, understand the tentative nature of SSI, and exhibit skepticism when analyzing information (Sadler et al., 2007).

Sadler et al. (2004) uncovered that in order for students to understand NOS, students must first understand what constitutes data and its potential uses. Their research findings revealed that students tend to believe that the most convincing position on a topic is the one already closely related to their own beliefs. The second finding revealed that students were drawn to a position because it presents consequences directly related to them. As a result, researchers urge educators to challenge students by providing opportunities for reflection, discourse, and integration of scientific knowledge, while evaluating alternative views to align with scientific consensus (Sadler et al., 2004). Consider engaging students in the activities listed in Exhibit 4.2 to develop their socioscientific reasoning skills.

Exhibit 4.2 Opportunities for Socioscientific Reasoning

To further expose students to NOS principles, try implementing the following activities and lesson plans with students to build on their science and engineering practices. When teachers present more than one perspective to a problem, they increase students' critical thinking capacity by complicating their beliefs through rigorous instruction.

1. *Two Sides Activity*—Present the class with fictitious science briefings summarizing opposing positions. Half the class receives one briefing showing evidence by scientists who report on climate change as anthropogenic and a real threat, while the other half is presented evidence suggesting that climate change is a natural phenomenon and not at all an environmental threat. Have them gather data to support the claims made by the *briefings* (not their own claims leading to a personal debate), and then play the role of fact finding detectives when they realize that both sides have "evidence" to support their scientists' claims. This activity also works when presenting opposing positions on

the benefits/dangers of Dihydrogen Monoxide (a.k.a. H_2O). Teachers can also ask students to sign an official petition to ban the substance before the big reveal (www.Bit.ly/DHMOban).

2. *Global Oneness Project Video Clips and Lesson Plans—*Consider having students learn about human impacts alongside climate science to frame your class discussions using these resources. When teachers have students engaging in argumentation over alternative sources of energy, basic human rights (i.e. access to clean air, food, water, etc.), or the degree of harm caused by climate change, students can engage in critical thinking by also considering stories told by real people impacted across the world. For example, when students consider the personal benefits of using fossil fuels, will they still feel that way when they see that it's at the expense of nature and other people? Students can also participate in video and photography contests to tell their own stories at (globalonenessproject.org).

3. *NYTimes Climate Change Lesson Plan—*Consider using the following lesson to learn about anthropogenic climate change through data analysis, impacts on people across the globe, and the complex nature of the climate crisis for critical discourse (www.Visit Bit.ly/NYTLP). Help students to make sense of climate change data and to create an iterative perspective of the climate crisis as they learn more throughout the storylines.

4. *Environmental Graphiti Art—*Alisa Singer is incredible at taking climate change graphic data and turning them into incredible pieces of art to present creative opportunities for people to engage. Her art campaigns can be seen across exhibits and the cities to support questioning, conversation, and to spark continued interest. Consider having students review graphic data and express that information in meaningful ways to them. Check out Alisa's virtual exhibit

(www.Bit.ly/ENVIROART) on climate change to inspire students to engage in critical thinking through art interpretation at the intersection of scientific findings. Can students figure out what Alisa is trying to convey through her work? What would they do to add to what they're seeing? What would they change to reflect their own thoughts and feelings about the data?

Taking More Back to the Class

Now that you know the benefits of using climate change topics to advance students' scientific literacy skills, it's time to take more back to class. Recall that the SSI Framework stresses the importance of *building instruction around a compelling issue* and *presenting it first* to provide deep context for learning (Presley et al., 2013). These real-world contexts provide authentic experiences allowing students to have more depth of knowledge with skills to take action on issues outside the classroom (Hammond, 2015; Sadler, 2009). The following segments will help you learn more about phenomena, how to find the strong ones, and how it can all circle back to the climate crisis.

Using Phenomena

A phenomenon is something observable that happens in the world that is not easily explained. There are two main types of phenomena (Anchoring and Investigative) you can include to help students make sense of what they are learning. Think of anchoring phenomena (AP) as big complex ideas, events, or questions that you can connect your everyday lessons back to. They are so complex that they involve many topics and concepts to fully grasp. The AP could take many weeks to fully understand and require several lessons, activities, lab experiences, discussions, etc., to fully unveil. The AP should also be open-ended and complex enough that every student can draw from a variety of resources to organize evidence needed to support their claims. Refer to Exhibit 4.3 for guidance on how to select strong AP for your storylines.

Exhibit 4.3 Finding Phenomena Criteria

Below are qualities to consider when you are selecting anchoring phenomena (AP):

1. A strong AP takes into consideration students, their communities, and their lived experiences. The anchor should be compelling to students in that they want to know more about what they are seeing or learning about. Take special consideration for underrepresented minority groups in STEM, your multilingual language learners, etc. Remember to think about the anchor serving as both "windows" into learning about and valuing other cultures, as well as "mirrors" where students see how they are represented and carry a presence in that lesson.

2. A strong AP requires students to engage in three-dimensional learning (DCI, CCC, SEP) to apply what they've learned to NGSS performance expectations. This process needs to be cyclical in that lessons intentionally connect back to the AP and are revisited at different points across instructional segments. It should also be iterative in that the sequence builds from one day to the next and pushes students to engage in a necessary productive struggle.

3. A strong AP should be big enough for students to explore across several weeks because it requires more than one idea, concept, approach, or solution.

   a. Allow students to develop a class driving question from observing or learning about the AP. This allows the claims made by individual students to be multifaceted and dependent on the evidence they collect over several lessons to use in their explanations.

b. Make sure to let students know that evidence collected must be from class lessons, activities, labs, discussions, explorations, etc. to support their claim for explaining the anchoring phenomenon and driving question.

c. Deliberately call out the NOS or scientific literacy skills needed to fully construct explanations for their claims (ex. Students should be encouraged to modify their claims in light of new information or evidence over time).

4. A strong AP is observable. Students can observe the AP through video clips that don't reveal too much information, pictures, a class demonstration, a societal problem (i.e. invasive species, coral bleaching, looking at drought data, etc.), the context for your lab experiment, etc.

a. The AP should make students want to ask questions about what they are seeing. Furthermore, it should make them want to learn more about it to give science context.

b. Students should be engaging in the science and engineering practices, crosscutting concepts, and applying relevant disciplinary core ideas across connected lessons in an iterative fashion.

5. A strong AP is important, relevant, and matters to students. It doesn't have to be phenomenal, but students should feel invested in learning more about what is happening to either create solutions for it, or take action when they have deep content knowledge on it.

As you think about what phenomena to use to anchor your instruction, note that the climate crisis fulfills all of the qualities above recommended by researchers and the NGSS.

Now that you know more about AP, let's explore supporting investigative phenomena (IP) and their role in storylines. Investigative phenomena are more direct ideas, events, or questions related (directly or indirectly) to the anchoring phenomenon. Investigative phenomena are used to provide additional information or clues to help unveil more about the main anchoring phenomenon. The answer to the IP could be uncovered the same day you introduce it and you could have an IP for each lesson, activity, or lab so long as it clearly connects back to the anchoring phenomenon. It is crucial that teachers provide time for students to make sense of how their daily lessons connect back to the AP because this is how students can engage in sense-making. Figure 4.1 shows an example of how to start a storyline with an anchoring phenomenon, supporting investigative phenomena, and anticipated student driving questions.

Collaborating and Designing With Students in Mind

As you begin to think about grounding curriculum on climate change topics for students, it will be tempting to use curricula developed by other schools or education companies. Having these starting points can be great, so long as you adapt it to the needs of your students with relevant phenomena they can take action on. Although that might serve as a great starting point, your lessons need to be driven by your students and allow for co-construction of knowledge through iterative sense-making opportunities. Remember that our underlying beliefs about teaching and learning directly impact our teaching decisions. What might be unacceptable to us, may be the best thing since sliced bread for another teacher because of their differing underlying beliefs about teaching. Don't forget to think about what you value as an educator as you create or comb through other people's lessons to ensure that you are proud of what you're teaching, that it reflects the current research embedded in the framework on how students learn science effectively, and that it aligns to your vision of science education to meet the needs of every student.

Teaching to spur climate action requires a culturally relevant and responsive approach that is locally driven and globally connected. As you narrow your focus on creating or adapting

NGSS Curriculum Development Storyline Tool

Anchoring Phenomenon	*Description:* Students will watch a clip from Chasing Coral showing the New Caledonia Reef undergoing a bleaching event and causing the coral to fluoresce. *Photo by Bawah Reserve on Unsplash*	*Connection to my students' interests, lived experiences, and funds of knowledge:* Students will be intrigued by what they are seeing because it is highly unusual. They also generally care for living things and many of my students live near the ocean. As they discovered more about anthropogenic factors, they will become more interested in how they directly contribute to such events that are seemingly far away.
The Driving or Anchoring Class Question (*Will be answered at the end with evidence to explain the phenomenon*)	*I anticipate my class will generate one of the following questions:* 1) What is causing the coral to change colors? 2) Why are the corals fluorescing? 3) What are the factors that lead to the corals producing the chemical sunscreen?	*Learning this will help to address or solve a personal, family, or community issue by...* I plan to use investigative phenomenon to teach students about their direct role and actions that lead to coral bleaching events. Also, I plan to show students how human health is reliant on the health of the ocean.
Potential Investigative Phenomena to Consider (*Must explicitly connect to the AP and require students to collect data and evidence for*)	*Layered strategically throughout the storyline:* 1) Watching a clip of faucet water catching on fire. 2) Watching a clip of river water on fire. 3) Looking at images of species that fell into Lake Natron in Tanzania. 4) Water and carbon cycle to understand how fossil fuels are created and role of limestone in the ocean. 5) Analyze data across the globe from NOAA on mass coral bleaching events. Ask more questions about the AP and possibly modify the AP based on data. 6) Reflect on human activities that would directly lead to an increase in CO₂ and change in pH in the ocean. 7) "Breath in a cup activity" 8) Acid-Base Titrations Group Inquiry Lab 9) Watching a short clip on ocean acidification to generate more questions. 10) Using science to regrow coral and help them become more resilient through music science.	*Connections to the AP:* 1) Students wonder about contaminants in water and how to filter it out. 2) Students learn more about what is in water and the role of how it impacts ecosystems and human health. 3) Students learn about potential of Hydrogen in water (pH), calculations for sense-making, and impacts on living things. 4) Students uncover more about human activities and CO₂ 5) Students analyze larger sets of data to see patterns to show larger scaled impacts beyond the New Caledonia Reef. 6) Students connect their direct role to the AP and think about changes that need and can be made at different levels. Students also consider the complex nature of the topic regarding benefits 7) Students analyze and collect data to see how CO₂ changes the pH of water by breathing into a cup of water with indicator. They explore ocean acidification. 8) Students engage in an inquiry lab where they are provided with various unidentified acids with varying molar concentrations that they have to titrate and neutralize. This leads to learning about buffers and how the ocean acts as a buffer as it tries to regulate Earth's climate to reach equilibrium. 9) Understanding large-scale impacts on the ocean that impacts marine species and entire ecosystems. 10) Using science and engineering as tools to develop solutions.

FIGURE 4.1 Starting Storylines with an Anchoring Phenomenon

resources, knowing your core values and beliefs about teaching and learning provides you with a stronger rationale and affirmation to make changes. To amplify the impact, it will be crucial to work with colleagues to determine the best curriculum for your students. Collaborating with colleagues should become easier because explaining *why* you want to advocate for certain components or *ways* of teaching to address equity issues (i.e. deliberate efforts to include discourse, start with meaningful phenomena, intentionally build in a culturally relevant lens, etc.) will be more clear. If a colleague disagrees with those lesson components or approaches, you will be able to listen with intent to help surface their values and beliefs about teaching and learning to have more productive conversations. When you're able to talk about the real challenges preventing your team from moving forward (i.e. disconnect in teacher values and beliefs about how or what students should learn in science, lack of shared group understanding and vision, the purpose of science education, etc.), you'll be able to have more effective solutions moving forward. As you begin creating storylines for your students, consider the following questions to guide the process:

1. What are initiatives prioritized at your school (student talk, writing across the curriculum, focus learning targets, discourse, literacy, etc.)?
2. What are your science department planning priorities this year? Climate can be the vehicle, but three-dimensional learning of NGSS is cross discipline so what potential levers can you pull on so it's not seen as an additional task?
3. Looking at the NGSS state exam benchmark scores for your school, what areas of improvement were identified in the test? Can increasing scientific literacy through integrating complex topics address those identified gaps?
4. What are your subject matter collaboration efforts like (if any)?
5. Revisit your vision of good science teaching in Chapter 1? What might that look, sound, and feel like for students? What are common values or beliefs shared among you and other subject matter teachers that you can collaborate with?

6. How would you describe the community you teach in? What are some of their needs that you've identified? Can schools be positioned to provide spaces to learn about and tackle some of those challenges?
7. What are the cultural backgrounds of your students? How do you learn about their deep cultural values (ex. Views of right versus wrong, how knowledge is passed down generationally, ideas about individualism versus collectivism, etc.)?
8. How are students typically positioned in class? In some of your colleagues' classes? What are similarities and differences you've noticed? Why do you think that?

After you have answered the questions above, consider engaging in Exhibit 4.4 to begin the process of creating new storylines centering on your students, school, and community.

Exhibit 4.4 Is This the Best Phenomenon?

 Launching the unit is key, and selecting the best anchor to drive the lesson is critical to developing students' complex critical thinking and socioscientific reasoning skills. Think about one phenomenon in one lesson that you are possibly interested in using for your lesson by answering the following questions:

♦ Is your phenomenon an anchoring or investigative phenomenon? Why?
♦ How excited are you about your chosen phenomena?
♦ What are the main science concepts students need to learn to understand the anchoring phenomenon?
♦ Through what methods will they learn about those science concepts?
♦ Will they be exploring through activities and/or labs at any point?
♦ When will they have opportunities for discourse to make sense of what they are learning?
♦ When will they engage in argumentation using evidence to support their claim?

◆ When and how will they learn NOS principles?
◆ Are there other phenomena you almost used for this unit that had more straightforward answers? Consider using these as investigative phenomena.
◆ Where can the investigative phenomena be placed throughout the storyline to continue engaging your students and provide more information about the Anchoring phenomenon?

Answering these questions will allow you to think about designing seamless storylines that are directly relevant to your students, their communities, and their interests.

 *Connect with the network to see how other teachers are integrating phenomenon-based instruction using at www.EmpoweredScienceTeachers.com (Book Resources → Chapter 4 → Discussion Board)*

Finding Phenomena

Consider the following resources in Table 4.2 to start your search. It can be overwhelming and not often helpful to search randomly, so remember to look with some goals in mind. Which phenomena are complex enough to create storylines around, that also sequence and build on depth of content and skills needed over time. Also consider potential driving questions that you anticipate students developing through the selected phenomena to guide their learning journey (this can also help you find relevant and/or student-driven investigative phenomena to integrate). It might be helpful to test out different phenomena with different class periods to see which one produces more aligned driving questions that can connect big ideas and topics. Given climate change's complex nature, anchoring instructional segments around the climate crisis allows for both cyclical, iterative, and relevant instruction to take place more fluidly. Look ahead at

TABLE 4.2 Phenomena Resources

Resources	Description
Access to Dr. Lê's Padlet	Explore a variety of resources to integrate climate change into your curriculum. There are always updated current events that make for great phenomena.
	www.Bit.ly/LePadlet *Password:* GCCEducator
Explore Phenomena by Grade	Search by grade level and standards for general phenomena.
	thewonderofscience.com/phenomenal
Ocean Phenomena Teaching Resources	Search by topic or secondary grades to gain access to current ocean related phenomena.
	theoceanagency.org/ocean-phenomena
Phenomena for NGSS	Look at a compilation of video clips to use as phenomena.
	www.ngssphenomena.com
Searching Phenomena by Topic	Type in a word search and by grade level to explore phenomena for your storylines.
	www.georgiascienceteacher.org

Table 4.3, which lists several climate change phenomena currently used in science storylines to help inspire your design process.

Anchoring Instruction on Climate Change Phenomena

Climate change content makes for ideal phenomena because they are complex and require many different topics and ideas to fully understand. This issue *directly* impacts students across *every* community, and is intersectional which demands diverse solutions. The undeniable disproportionate impacts felt now and into the future by students also opens up opportunities to engage in discourse, deep empathy building, and emotional resilience to center on hope and justice. As students are driven to take action on environmental and climate issues, we can support them by providing opportunities to learn about the complex nature of climate change. Approaching climate science as an SSI is how educators

TABLE 4.3 List of Climate Change Phenomena

Phenomena Ideas	Link (case-sensitive)
Boiling seas (methane trapped in ice)	www.Bit.ly/BoilingSeas
CA environmental impacts and worsening human health impacts	www.Bit.ly/CALENVIRO
COVID-19 Pandemic (Climate change and infectious diseases)	www.Bit.ly/covidngcc
Darvaza gas crater (release of methane gas since the 1950s)	www.Bit.ly/DARVAZA
Diver in Bali documenting plastic waste	www.Bit.ly/BaliDiver
Emerging "Hunger Stones" in Germany	www.Bit.ly/HungerStones
Extinction of a species in the wild (Death of the last male northern white rhino).	www.Bit.ly/SudanPic
Faucet water that catches on fire due to fracking	www.Bit.ly/Wateronfire
Florida "raining" iguanas	www.Bit.ly/IguanasFL
Ghost Forests (Impact on trees)	www.Bit.ly/GhostForest
Harvesting drinking water from thin air	www.Bit.ly/waterinair
Looking at water runoff and snowmelt feedback effects	www.Bit.ly/ioeswater
Microplastics found in human blood	www.Bit.ly/PLASTICBL
Migrating spiders from Mexico entering California	www.Bit.ly/spidersew
MRSA or CRE outbreaks (Antibiotic resistance)	www.Bit.ly/mrsaspread
Mysterious balls of goo washing ashore (stranded jellyfish)	www.Bit.ly/JFgoo
Ocean dead zones (agricultural runoff and human activities)	www.Bit.ly/AGdeadzone
Sea Stars that rip their own limbs apart (wasting disease)	www.Bit.ly/SSspecies
Silicon Valley invasive species	www.Bit.ly/SVinvasive
The sixth-mass extinction	www.Bit.ly/6MExtinct
Traces of microplastic circulating in the Arctic and Antarctic	www.Bit.ly/Microplastic
Tracking forever chemicals (PFAs) across the USA	www.Bit.ly/PFACHEM
Urban Heat Island Effect (Self-reinforcing feedback loop)	www.Bit.ly/HeatEffect
World-wide coral bleaching events (start with Chasing Coral)	www.Bit.ly/WWbleaching

can begin to use their power and privileges to help students think about the relevant solutions needed, and opportunities for entering the green workforce with critical skills to thrive.

Connecting the Dots

Refer back to the curriculum map that you created in Chapter 3. As you think about integrating climate change phenomena into your units, you can add those AP or IP to relevant content topics. This might look like the following:

◆ Adding ocean acidification as a phenomenon that you can add to the unit to teach about acids and bases beyond the basic chemistry,

◆ The melting permafrost waking up ancient viruses as a phenomenon to your lesson on evolution with the context that this is happening today and the importance of understanding viruses as both superheroes (used to treat patients) and villains (what they normally associate viruses with),

◆ Integrating Earth's energy budget to understand solar energy to learn about waves, electromagnetic radiation, or conservation of energy,

◆ Role-playing decarbonization efforts as integrated in physics to learn about different forms of energy currently employed and where in the world, to apply nuclear energy concepts,

◆ Pointing out the reliance of consumerism with everyday plastics, but through a systems-thinking lens by asking students how studies are currently revealing the microplastics both in every breath we take and also now present in human blood.

Once you feel that you have a handful of phenomena for that instructional segment, you can begin to piece your storyline together. This chapter introduces you to phenomena-based instruction with resources to help you focus your search. When you're ready to design whole storylines centered on climate change, Chapter 5 will provide you with teacher tools and resources to support the design process. Continue adding to your own curriculum map (see Appendix B) or refer to Appendix C for different curriculum maps created by science teachers thinking through big ideas grounded on the climate crisis.

 *Consider submitting a Flipgrid video to the network for feedback or ideas. You can talk through your ideas, how you anticipate students will experience the phenomenon, and what you're looking to improve or gain ideas for at flipgrid. com/teachclimatechange (Password: Empowered).*

Track your professional growth by re-examining your beliefs about teaching and learning.

Tracking Your Professional Progress

Student Learning Experiences in Your Class	On a scale of 1 (extremely disagree) to 5 (extremely agree), to what extent do I agree or disagree with the following statements? Why?
Today's Date:	
Students often discuss policies related to science.	
Students often collect and analyze current data or information.	
Students often discuss ethical issues related to science.	
Students often co-construct knowledge with me during lessons.	
Students often drive the instruction in my class as capable contributors and doers of science.	
Students are often positioned as current scientists or engineers.	
Students often learn about Indigenous traditional ecological knowledge.	

Curriculum Design Elements	On a scale of 1 (extremely disagree) to 5 (extremely agree), to what extent do I agree or disagree with the following statements? Why?
Today's Date:	
I often build lessons/units around anchoring or investigative phenomena.	
I often present climate or environmental issues at the start of each unit or lesson.	
Students often engage in argumentation and making claims based on evidence.	
Students often engage in meaningful discourse opportunities.	
My lessons/units are often centered around real-world issues that are directly related to students' lives or community.	
Students often engage with the Nature of Science principles.	
Students often connect with local or relevant field experts and researchers.	

My Teacher Attributes	On a scale of 1 (extremely disagree) to 5 (extremely agree), to what extent do I agree or disagree with the following statements? Why?
Today's Date:	
I have to know everything about a particular topic before teaching it.	
I am extremely confident in my knowledge of climate science.	
I feel comfortable admitting to students when I don't know the answer to their question.	

I am comfortable teaching about open-ended issues where I cannot predict student responses.	
I am not the only source of knowledge on climate and the environment for students.	
I often experience imposter syndrome when teaching, even for topics that I have strong expertise in.	
I often reflect on my pedagogical decisions to improve my practice.	

 *Refer back to the **Book Introduction** where you initially tracked your professional growth. Reflect on how your beliefs might have changed and why that might be. How might your responses unveil more about your teaching disposition?*

Collective Voices for Climate Change Education

Candice Dickens-Russell, CEO of Friends of Los Angeles River

In the realm of teaching about climate change, you stand as a beacon of inspiration, guiding the next generation toward environmental stewardship & justice. Today's students are not just learners; they are informed advocates, armed with a profound understanding of climate issues. As an educator, your role extends beyond the traditional classroom, evolving into a collaborative journey where students and teachers coalesce as partners in change. Acknowledge the profound wisdom your students bring, for they are not merely recipients of knowledge but active contributors to the discourse. In this era of heightened environmental awareness, you're not just leading; you're navigating alongside your students, fostering a space where their unique expertise can thrive. Embrace this collective responsibility, championing not only the importance of ecological sustainability but also the imperative of environmental justice. Your commitment to empowering students to become informed and compassionate stewards of our planet sets the stage for a future where both knowledge and action intertwine seamlessly, creating a legacy of positive change for generations to come.

Alisa Singer, Artist and Founder of Environmental Graffiti

I believe art can serve as a uniquely effective and powerful tool to communicate scientific information. Crossing all boundaries—language, geography, ideology, politics, ethics and religion—art invites people of all backgrounds into the room to begin the critical discussion about what is happening to our planet, and to the plant and animal species that share it with us. ART makes SCIENCE more accessible. SCIENCE makes ART more meaningful. It's a potent combination.

Additional Teacher Resources

Access ESRI's resources to teach about nature of the environment—
   Bit.ly/esriNATURE
Alternative Phenomena Criteria Checklist by Researchers—
   Bit.ly/PhenomChecklist
Basic Physics of Climate Change—
   Bit.ly/GCCPhysics
Energy Map L.E.A.D. Tool—
   energy.gov/scep/slsc/lead-tool

NYTimes resource to track how much hotter your hometown is now—
Bit.ly/hotterdata
Pin this Pinterest board by STEMeducation for Indigenous science teaching resources—
www.Bit.ly/TRpinterest
Preview data in the US impacting humans—
Bit.ly/Fractracker
Preview Project Drawdown's Climate 101 learning series—
drawdown.org/climate-solutions-101
Teaching Channel explains phenomena—
Bit.ly/TCPhenom
Use NOAA's Data in the Classroom—
https://dataintheclassroom.noaa.gov/
Read ASCD's article on teaching scientific literacy—
Bit.ly/TeachingSL
Read the full 100kin10 2019 Trends Report—
Bit.ly/100KREPORTS
Watch clips from Conservation International series—
conservation.org/nature-is-speaking/
Watch "Kiss the Ground" to see how people are combating climate change through agriculture—
https://kissthegroundmovie.com/
What Americans know and don't know about science—
Bit.ly/PEWRESULTS

References

2019 TRENDS REPORT: *Trends and predictions that will define STEM in 2020.* (n.d.). Retrieved May 13, 2020, from https://beyond100k.org/2019-trends-report-trends-and-predictions-that-will-define-stem-in-2020/

Anelli, C. (2011). Scientific literacy: What is it, are we teaching it, and does it matter. *American Entomologist*, 57(4), 235–244.

Caranto, B. F., & Pitpitunge, A. D. (2015). Students' knowledge on climate change: Implications on interdisciplinary learning. In *Biology education and research in a changing planet* (pp. 21–30). Singapore: Springer.

Carter, B. E., & Wiles, J. R. (2014). Scientific consensus and social controversy: Exploring relationships between students' conceptions of the nature of science, biological evolution, and global climate change. *Evolution: Education and Outreach*, 7(1), 6.

Hammond, Z. (2015). *Culturally responsive teaching and the brain: Promoting authentic engagement and rigor among culturally and linguistically diverse students*. Thousand Oaks, CA: Corwin.

Hansen, P. J. K. (2010). Knowledge about the greenhouse effect and the effects of the Ozone Layer among Norwegian pupils finishing compulsory education in 1989, 1993, and 2005—What now? *International Journal of Science Education*, 32(3), 397–419.

Hestness, E., McDonald, R. C., Breslyn, W., McGinnis, J. R., & Mouza, C. (2014). Science teacher professional development in climate change education informed by the next generation science standards. *Journal of Geoscience Education*, 62(3), 319–329.

Hodson, D. (2003). Time for action: Science education for an alternative future. *International Journal of Science Education*, 25(6), 645–670.

Kendi, I. X. (2019). *How to be an antiracist*. First Edition. New York: One World.

Khishfe, R., & Lederman, N. (2006). Teaching nature of science within a controversial topic: Integrated versus nonintegrated. *Journal of Research in Science Teaching*, 43(4), 395–418.

Kolstø, S. D. (2001). Scientific literacy for citizenship: Tools for dealing with the science dimension of controversial socioscientific issues. *Science Education*, 85(3), 291–310.

Matkins, J. J., & Bell, R. L. (2007). Awakening the scientist inside: Global climate change and the nature of science in an elementary science methods course. *Journal of Science Teacher Education*, 18(2), 137–163.

National Research Council. (1996). *National science education standards*. Washington, DC: The National Academies Press. https://doi.org/10.17226/4962.

Presley, M. L., Sickel, A. J., Muslu, N., Merle-Johnson, D., Witzig, S. B., Izci, K., & Sadler, T. D. (2013). A framework for socio-scientific issues based education. *Science Educator*, 22(1), 26.

Sadler, T. D. (2009). Situated learning in science education: Socio-scientific issues as contexts for practice. *Studies in Science Education*, *45*(1), 1–42.

Sadler, T. D., Barab, S. A., & Scott, B. (2007). What do students gain by engaging in socioscientific inquiry? *Research in Science Education*, 37(4), 371–391.

Sadler, T. D., Chambers, F. W., & Zeidler, D. L. (2004). Student conceptualizations of the nature of science in response to a socioscientific issue. *International Journal of Science Education*, 26(4), 387–409.

Somerville, R. C., & Hassol, S. J. (2011). The science of climate change. *Physics Today*, 64(10), 48.

Part 3

Practices That Build Capacity for Student Agency

5

Planning and Teaching for Transformation

Read this when:

♦ *You're ready to move beyond phenomena-based instruction to designing meaningful storylines.*

♦ *You need guidance on NGSS-aligned and evidence-based teaching practices to integrate with current or relevant storylines.*

♦ *You want to see more examples of lessons that position climate change as a socioscientific issue (SSI) to develop students' scientific literacy skills.*

Take a moment to learn about the following story which showcases innovation, persistence, and resilience in tackling a community energy and environmental waste issue. Carvey Ehren Maigue is an engineering student from the Philippines, who is currently developing renewable and sustainable energy using food waste. It started when he first noticed that the lens of his transition glasses would darken even on cloudy or rainy days. This led him to see how ultraviolet (UV) light is still able to reach him even when the Sun's rays are blocked by clouds. Maigue decided to find novel ways to capture and convert this energy source. Examining his own community and cultural practices, he experimented with food waste as a potential game changer—specifically locally

DOI: 10.4324/9781003478584-9

grown fruits and vegetables that contain organic luminescent compounds (they glow in the dark). Through numerous attempts, he discovered that these compounds are able to convert high energy UV waves into visible light. Using solar panels and films, he then turned this energy source into electricity. After many failed attempts and iterations, Maigue finally succeeded and his invention (Aurora Renewable Energy and UV Sequestration [AuREUS]) can now be used on buildings, homes, clothing, etc., to generate renewable electricity. Maigue went on to become the 2020 James Dyson Global Sustainability Award winner.

Nalleli Cobo is one of my personal (s)heros. If you haven't heard of this incredibly brave young woman from Los Angeles, California, look her up! Nalleli has been leading with courage and fearlessness starting at the age of nine when she brought the issues of environmental racism and injustice to her local school board. When she started suffering from severe health issues (like bloody noses so intense she had to sleep sitting up as to not choke on her own blood), they suspected that it might be due to the oil rig in their neighborhood. Oil rigs in Los Angeles can be found almost everywhere (but especially present in historically marginalized communities). The laxed regulations on oil rigs meant that homes, schools, buildings, parks, etc., could be built next to the rigs even at the expense of human health (an estimated 75% are located near sensitive areas typically used or occupied by people). When Nalleli took note of the toxic fumes and odors that came from the rigs, she and her family went door to door in their community to talk with neighbors to see if they were also experiencing similar health issues. Those community members knocked on additional doors and what they found was a pattern of chronic health issues from those living around the oil rig 30 feet from Nalleli's home. She and her family took matters into their own hands to seek justice for current and future members of the community to ensure that no one else has to experience this type of environmental racism and injustice.

The 2022 Goldman Prize Winner has since taken on big oil and successfully built coalitions to challenge California's regulations, which are now closing and or phasing out oil rigs due to new buffer zone requirements. As Dr. Nadia Kim unveils in numerous accounts from community members, there are many stories of people of color "refusing death" in the fight for environmental justice in Los Angeles (2021). Whether it is about oil rigs, refineries, lack of access to clean water, or excess emergency visits due to asthma attacks induced by bad air quality—students and their families recognize their differing lived realities. Could schools serve as organizations of change to serve as solution-oriented spaces and incubators of innovation so that students can apply what they learn to tackle real issues and challenges in their communities? What might that look like? How might we foster the skills, knowledge, and opportunities needed to create that type of space? What beliefs would people need to hold in order for this to be a reality?

We need to trust that people hold solutions to climate change in their own communities and remember that our students are capable of bold leadership. When we talk about student agency, we are referring to the capacity and ability to act that is already within them. With guidance, modeling, and allyship, we can support students to take meaningful action on science issues that directly impact them.

A Vision for Science Education

When you think of a complex problem and find yourself overwhelmed on how to start tackling it, start with the end goal and sequence instruction from that point (also known as backwards planning). Being explicit with students about the learning process will only help them to better understand your methodology, make your thinking visible, and get them to trust the process as well.

Backwards planning for gapless explanations is an approach often used to create NGSS storylines. Although previously referenced, this chapter will provide more depth regarding storylining for a shared understanding of the framework. Refer to the curriculum map you started in Chapter 3, and select *one* upcoming unit that you plan to teach. Then, access the storyline development tool online at *empoweredscienceteachers.com* → Book Resources. Write down any big ideas or concepts in the unit selected that are non-negotiables to set the parameters. Next, list *smaller* ideas and concepts needed to better understand the *larger* ideas or concepts you identified. Although you might have ideas on how to build out the storyline, consider the following questions:

1. Which phenomena are directly or indirectly connected to the major concepts of this unit?
   ◆ Which are culturally relevant or responsive? In what ways?
   ◆ Which allows for students to take action on the issues they're presented with?
2. What skills will students need to develop or learn to master the major concepts identified?
   ◆ At what point in the lesson can they build scientific literacy skills or develop socioscientific reasoning skills?
   ◆ How will they engage with SEPs or NOS principles? How often? To what depth?
3. When do students engage in the Process of Science (actually doing science)?
   ◆ Is the problem open-ended enough to allow for students to tap into their own cultural wealth and funds of knowledge to begin looking for solutions (refer back to Figure 1.1 and consider accessing the tool to support this)?
   ◆ Who is driving the lessons and how are they positioned as a result? How strong is the alignment to the process of how scientists, researchers, or engineers do work?

This chapter will help organize your thoughts around these questions with more application opportunities in design.

Overview of Storyline Design Components

Brian Reiser, a professor from Northwestern University who supports science teachers with NGSS, states that a storyline is a coherent sequence of lessons driven by students' questions that emerge from relevant phenomena. As you begin to revamp your curriculum, it's essential to see the larger picture first then backwards plan. Exhibit 5.1 provides an overview of what storylines are and how lessons or activities are sequenced.

Exhibit 5.1 Understanding Storylines

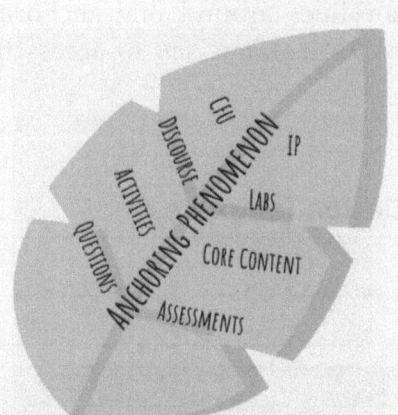

Imagine a storyline resembling parts of a tree leaf. When you look at a leaf, there is typically one central vein that starts from the bottom that may stretch all the way to the top. That central vein is your anchoring phenomenon (AP). When you look closer at this leaf, you start to notice smaller veins that break off from the central vein. Those smaller side veins are the components of your storyline (such as investigative phenomena (IP), CER, discourse, lab experiences, activities, core content, etc.). What teachers tend to miss when designing the curriculum is how the smaller veins must always connect back to the central vein. In short, all that you teach in the unit needs to directly connect back to the anchoring phenomenon, and this needs to be clear for students through reflection or discussion opportunities for sense-making. Those lesson components should provide students with more information and evidence needed to explain their driving question developed from the AP.

As you build more storylines, imagine looking at different leaves that sit on the same branch. Although the branch itself represents the science course, the different leaves represent different instructional segments that are all connected through various anchoring and investigative phenomena. Can you imagine designing science curricula as a firmly grounded tree that branches off in different directions (subject areas), but always connects back to the climate crisis (the most pressing issue of our time)? Climate change is one of the only topics that allows for this to happen successfully because the problem is complex and requires an intersectional approach to learn and create solutions for.

With the bigger picture in mind, it's time to explore specific steps to bring this curriculum to life. Figure 5.1 provides a visual to help you design storylines around one anchoring phenomenon (Visit *empoweredscienceteachers.com* to access the online template).

Storyline Unit Design Template

Instructional Segment		Course	
Semester		Weeks & # of Days Needed	

Setting the Stage: Storyline Overview
NGSS Performance Expectations:
Anchoring Phenomenon:
Reflect: Is it culturally relevant, meaningful (towards students or their community), allows my students to take action or apply science in some way, compelling, allows for students to engage in the SEPs to explore the topic's ethics, etc.?
Students already know and are capable of the following:
Anticipated Student Questions & Initial Driving Question:
Anticipated Student Questions - Three questions I anticipate students posing from watching the clip of the AP are: 1) 2) 3) **Initial Class Driving Question -**
NGSS Three-Dimensional Teaching Components (DCI, CCC, SEP):
Other Standards Addressed:
NGSS Nature of Science: CA Environmental Principles & Concepts (EP&Cs): Climate Literacy Standards: Common Core Math/English:

| Anchoring Phenom | Investigative Phenom | Investigative Phenom | Investigative Phenom | Summative Assessment |

FIGURE 5.1 Storyline Planning Tool

Lesson Checkpoints: Sequence & Flow			
Core Content and/or Anchoring or Investigative Phenomenon	What will students be doing?	Why are students doing this and how does it relate to the AP?	Where will students go next to build on their knowledge?
AP: Core Content Knowledge:			
Vocabulary words (based on core content of the day): -			
Student Self-Assessment - How will students know they are on target to move forward? What will they need to know?	1. 2. 3.		
Vocabulary words (based on core content of the day): -			
Student Self-Assessment - How will students know they are on target to move forward? What will they need to know?			
Vocabulary words (based on core content of the day): -			
Student Self-Assessment - How will students know they are on target to move forward? What will they need to know?			

Application of Knowledge: Performance Task
This could be an end of unit project, video, work of art, pre-post modeling, answering the driving question with evidence/data/information gathered each day and showing the modified claims, telling a story through pictures to apply what they learned (Taking action in the community in some way), inquiry lab experiment, etc.

FIGURE 5.1 (*continued*)

Newmann, Smith, Allensworth, and Bryk (2001) argue that students are more likely to learn new content when experiences build on one another. They also note that content learned through short-term exposure and referenced minimally will not be retained or transferred to other settings. When we think about how students learn, this does not come as a surprise to any skilled teacher. That is why the NGSS requires a cyclical approach to allow students to revisit core concepts often as they build in complexity over time. If you look back at Figure 5.1, each lesson in the storyline relates to a larger driving question developed from the anchoring phenomenon to ensure that concepts are consistently cycling through the storyline. This iterative process allows for students to see and reflect on their own thinking as they gather data and evidence to explain the anchoring phenomenon.

Components of Strong Supporting Lessons

Let's address the daily lessons that are part of the storyline to ensure that you're not only teaching the words written in the NGSS, but that your pedagogical practices are also aligned to what the framework calls for. Recall that students learn when (1) teachers address their current understandings and cultural funds of knowledge, (2) they understand the true Process and Nature of Science and are positioned as capable doers and knowers of science, and (3) they are given opportunities to be metacognitive for lessons that serve as both "windows" and "mirrors" grounded in equity and antiracist teaching practices (NRC, 2005, 2018). How might we teach science in ways that position students as the drivers of culturally relevant and responsive lessons? Like anything new, it will take time and practice to integrate all three elements into your daily lessons. More importantly, think about whether these research-based approaches align with your values as an educator (see Exhibit 5.2 for an example of what this might look and feel like). When you see every student engage in the productive struggle (Hammond, 2014) and build on their science identity, confidence, and agency over time, you will find yourself not wanting to teach any other way.

Exhibit 5.2 Matching the Pedagogy to the Design

The following is a breakdown of a lesson plan to show its skeletal components. This lesson takes place on the first day of a new storyline.

1. Start by going over focus learning targets for the day on the whiteboard (can be applying SEPs or NOS principles).
2. Introduce the climate change-related anchoring phenomenon using a video clip/picture/demo.
3. Facilitate class development of the driving question for the unit using students' personal questions or wonderings.
4. Make it clear to all students that the class' driving question serves as their summative essay question at the end of the unit in x weeks. This will help them choose a big enough question to gather evidence for, reflect on their learning each day to see how it connects to the AP, and be in control of their grade.
5. Groups of 3–4 students share their personal questions to each other and then select one as their favorite group question. After writing all the questions posed by each group, the students vote for the one that is the strongest (the one they feel confident gathering for throughout the entire unit).
6. Then have them share initial thoughts and ideas with an elbow partner.
7. The teacher finally unveils the title of the unit to start the first lesson in the storyline.
8. The teacher continues to introduce the first big idea with visuals while posing sense-making questions.
9. Students engage in small and whole group discourse every 5–7 minutes during the lesson where they pose sense-making questions to each other and try to figure out how the information relates to the driving question.
10. The teacher stops at slide 5 and tells students to talk with a partner about what they are learning so far and how it may serve as evidence to support, refute, or modify their initial claim for the driving question.
11. Students return to the anchoring question page, and add evidence and re-evaluate their initial thoughts.

12. The teacher continues with the lesson by doing a short demonstration to showcase how the big idea works.
13. Students are prompted to return to the anchoring question page again to modify or add ideas or evidence.
14. The lesson closes by having students share with each other their current claims (which might have changed in light of new information), and evidence they gathered to support, refute, or modify their initial claim.
15. The teacher facilitates whole group discourse to hear students' ideas and understanding of content.

Although this is not the only way to begin a new storyline, it allows for you to see components of the lesson that address all three learning elements. Please identify which parts of the lesson (a) unveils students' current understandings, (b) allows for them to learn about the nature or process of science, and (c) pushes students to be metacognitive.

 *Consider answering the following questions to see where you are in your lesson planning, and what you consider to be essential in design.*

◆ What was or was not surprising about the launch example?
◆ Which learning element(s) do you already engage students with?
◆ How did this position students as co-constructors of knowledge with the teacher?
◆ What are some learning elements that you wish to strengthen for students after reflecting?
◆ How do you know when students' thinking has grown or changed? How does this launch help them to see that for themselves?
◆ When do they engage with NOS practices in this launch?

 *Looking at my answers and current teaching beliefs, my teaching goals are...*

 *These are components that you can start integrating into your lesson now. Take it back to the class and note how your class sounds, feels, and looks like when students engage in learning science with these elements. Which elements did you integrate and which did you leave out? Why?*

Student-Driven Instruction

Remember that the NGSS requires a cyclical and iterative approach. Although teachers may undergo various NGSS training, this central idea may never surface unless it is explicitly called out. Having a driving question allows teachers to revisit central ideas across lessons and instructional segments more easily. Giving students the opportunity to develop that driving question from the anchoring phenomenon is even more powerful. As many program teachers have shared with me, allowing students to generate the anchoring question for the storyline provided them with purpose for what they are learning. It also encourages students to make connections across lessons as they unveil evidence needed to make sense of the anchoring phenomenon. The level of engagement heightens for students that take ownership of their learning, and they become more invested. From the very beginning of the storyline, students are positioned as capable contributors whose ideas and funds of knowledge are driving the lesson. With this approach, teachers are able to provide three-dimensional learning for NGSS through a culturally relevant and responsive lens.

Launching Lessons with Phenomena

You can successfully facilitate the launch of the lesson with students as drivers in ways that are not "chaotic" or time-consuming. Think back to the phenomenon you chose in Chapter 4 as the starting point. Another possible starting point could be the driving question(s) dictated by your school district curriculum map. Select a video clip, image, or demonstration you feel would help students to develop the driving question you're aiming for. To see what types of students will generate, practice by showing the phenomenon to non-science friends or colleagues and ask them to provide you with three questions they have after viewing the clip, image, or demonstration. If one of the questions provided was close to the driving question you're aiming for, that is an early strong indicator for what to expect from students. Take the risk and put students in the driver seat to build their capacity as innovative young leaders. This small but meaningful exercise will allow for teachers to gauge what students know, how much they know, what they want to know more of, and how far we can push them to reach their potential.

Approaches to Support Students to Develop the Driving Questions

These are evidence-based approaches currently used by teachers to position students as capable contributors in the classroom.

First look. To introduce students to this approach, provide them with the driving question worksheet (see Appendix D) and post-it notes if possible. Show them a short (1–3 minute) clip or picture of a phenomenon without giving away too much context. Ask them to think about and write down one question they have about what they are seeing. Let the students know that they need to select one question as a group that, if the teacher provided the complete answer for it, would explain everything about that phenomenon. After 60 seconds, assign small groups of 3–4 and have them all share their individual questions. After giving each group two minutes to decide, go around the room and have each group share their chosen question out loud. During this time, support the entire class by having them write all the group questions being shared into the worksheet. Lastly, decide as a class by taking quick votes, on which question will be the class' driving question (see Figure 5.2 for an example). Remind students that the question chosen should be big enough to collect evidence for every day, and that they are deciding the essay question for their summative assessment. Troubleshoot issues that may arise by referring to Exhibit 5.3.

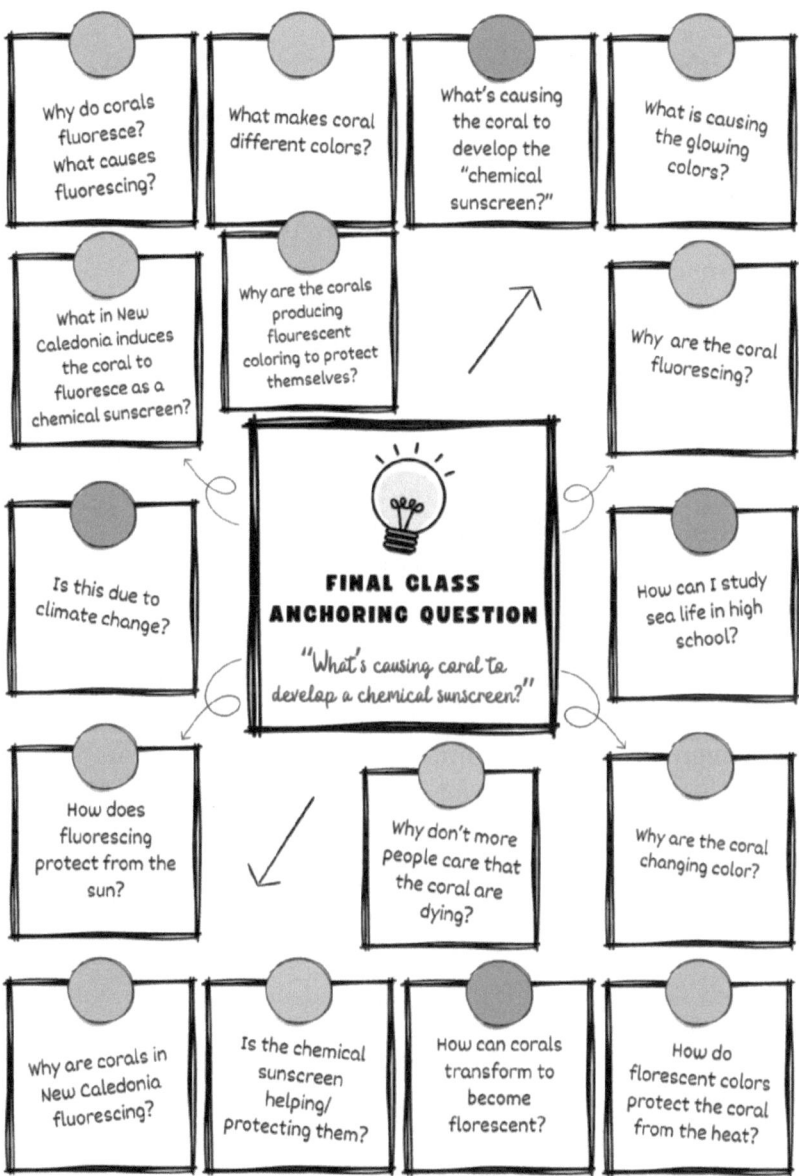

FIGURE 5.2 Generating Student Driving Question Samples

Exhibit 5.3 Tips for Supporting Student-Driven Questions

The following will help you troubleshoot common occurrences with these teaching approaches.

Question: What happens if different periods generate different questions?
Answer: Yes that might happen! Over the years I have learned that each class varies slightly in wording, but generally they are all asking and wondering the same thing.

Question: How much information do you provide when you show the video clip, image, or do the demonstration?
Answer: I give students only basic information in the event they don't know what to focus on, or if they don't know what something is during viewing.

Question: Are you providing them with any other information during that time?
Answer: Every student must write one question down on the post-it note or worksheet during viewing in roughly 60 seconds. To support them as they are viewing, I typically say one of the following multiple times as I circulate around the room:

"What is something you're currently wondering about?"
"What is a question that you have about what you're watching?"

When I group them together to use their questions to develop the class' anchoring question, I guide them further by saying multiple times, "Don't forget, look at the questions from your group. You can only pick *one* to share with other groups so pick the one that **if I answered it fully it would explain everything about the phenomenon.**"

Question: What happens if groups provide the same question when shared out loud?

Answer: I ask the group to share a different question presented to give the class a variety to choose from. I also don't tell students which question is the desired question to remain objective to ensure they have more voice and choice starting from the beginning of the unit.

Question: My class did not generate the driving question that I anticipated, what do I do?
Answer: Engage in the process of science as a teacher. So you didn't pick the right clip and their question wasn't exactly what you wanted, that tells you to either pick a different clip or adjust your facilitation of question generation with students (think about how you're guiding them toward sense-making to help them develop that skill that you already have). Make sure to have two different phenomena options to use in case the first one completely disappoints and reflect on your process for improvement.

Question: What if one of the group's poses a "yes" or "no" question, or they pose a question that is actually two different questions together.
Answer: The first time you model for students which questions don't work well (with intentional care to not embarrass any group). You can look at the group results and have students actually ask you the closed-ended question out loud. It will play out something like the following scenario:

Teacher—"Okay class so, let's look at group 4's question. The question reads, 'is the coral changing colors because it's dying?' Someone go ahead and pose that question to me."
Student—"Is the coral changing colors because it's dying?"
Teacher—"Yes." (silent pause). "Now that I have answered your question, does that tell you **why** the phenomenon is happening?"
Students—"No."

Teacher—"So that's why we don't want to select a closed-ended question as our anchoring question because I need to be able to build a case over time using evidence. I can't do that with a yes/no question. No worries everyone, we're all learning. How about I give every group 60 seconds to rewrite their question in case you have a closed-ended question."

Question: How long does the process of generating the driving question typically take?

Answer: The very first time generally takes 25 minutes because you have to take time to explain what they are doing and what a phenomenon is. Beginning the second time, it should be around 15 minutes before jumping right into the content because they understand the procedures of the activity.

Question: What if a student says they've seen the phenomenon and can look up the answer online?

Answer: Set ground rules at the beginning letting them know that they are investigators collecting evidence provided in class through lessons, labs, activities, discussion, etc. **only**. The point is to engage in the process of science and engage in argumentation using credible evidence provided from the class. Also, if they're researching more outside of class that's a win! That means they're engaged and can share more during class in ways that might influence your next investigative phenomena in the storyline.

Second look. To continue engaging students in this iterative process, consider this second approach when students are comfortable for any AP or IP. Show students a picture of a phenomenon (like a picture of Sudan who was the last male Northern White Rhino, with his caretaker moments before the animal died and his species was pronounced extinct in 2018, even though there were still two living females to push students to think about why the declaration was made). Have students generate a question based on what they are wondering about, and repeat the process from the "First look." After the driving question is determined, have students write down what they think the answer is for the phenomenon (this is their initial claim). If appropriate, teachers can also use a mental modeling worksheet that allows for students to draw their initial ideas. Have students add or modify their model every day given what they learned.

Third look. Once students begin to anticipate the type of question they should ask and select as a class (they will develop this critical skill pretty quickly). Challenge them by showing two clips of the same phenomenon that are slightly different to push them to modify their driving question to satisfy both events. For example, teachers can show a clip of a river on fire in Southwest Queensland to generate a class driving question, then show a second clip where a woman in the United States shows how the water from her sink catches on fire. Students are tasked to modify their question to ensure it addresses both phenomena prior to selecting a class driving question. This is also an opportunity for students to ask questions about phenomena that occur in different places that are similar to think through a systems thinking lens.

Fourth look. To get students to critically analyze questions, this next approach calls for them to go through the same process while using criteria to evaluate their group question prior to sharing it out loud. The criteria in Exhibit 5.4 can be provided to all students to help develop higher-level questions that call for more critical thinking. It can be used with any phenomena and with the same teaching approaches.

Exhibit 5.4 Criteria List for Students to Build Capacity

Students can use the following criteria as a checklist to strengthen their driving question.

◆ *It's Complex*—Is the question a yes or no question? If so, change it! What do you want to know that **if I answered it would explain the entire phenomenon?**

◆ *Definitely Relevant*—Does your question relate to something in the real world (man-made or natural event)? Can it be something that potentially relates to you or your community directly or indirectly? If yes, you're on the right track!

◆ *Open-Ended*—The question should not be too specific, understand that there are many paths to get to the final answer. The question is big enough that what you learn each day can help you learn a little bit more. Remember that some answers will be stronger than others depending on supporting evidence and credibility of evidence.

◆ *Take a Side*—The question should allow for you to make a claim and argue your position using credible sources of evidence from class.

◆ *Fascinating*—The question is interesting and seen as a mystery to you. You want to know more about this phenomenon and it makes you want to ask even more questions!

◆ *Scientific*—The question can only be answered by doing science and engineering (these are known as the Process and Nature of Science).

Consider submitting a Flipgrid video to the network for feedback or ideas on your storyline or lesson at flipgrid. com/teachclimatechange *(Password: Empowered).*

Intentional Discourse Opportunities

Although providing discourse opportunities is not a novel idea, the following are tools to help make the process more consistent and meaningful for students. Recall that the first lesson design component that fulfills both NGSS and Student Learning Elements is *unveiling students' current understandings* and *funds of knowledge*. Context is needed to bring meaning to the content, and students need to make sense of what they are learning and why this lesson matters. As educators, if we're having difficulty seeing how the lessons relate to/or honors students' lived realities along with addressing 21st-century problems, it will be even more confusing for our students to know the connections. Are we teaching what students need to know to address the problems of today?

Exhibit 5.5 Opportunities to Reflect and Apply

Consider answering the following questions to see where you are in your curriculum design, and what you're identifying now as "non-negotiables" in design.

- What are your thoughts on the current state of climate education? What are you happy with and what are you currently questioning?
- In what ways will climate change education support educators to revamp current science content to be more culturally relevant to place and people?
- How will you gauge the depth of knowledge by all students and how might you support them to gauge their own learning progressions (building metacognition)?
- How will you help develop their critical thinking skills? What might this look like in the curriculum and how often is this needed to support their growth?
- How might intentional discourse between students and with you develop their written and/or oral literacy skills? How might you lean into their funds of knowledge to strengthen this?
- When designing a storyline and the lessons that support it, how much will the students co-construct knowledge with you? When will they need your guidance and facilitation more? When can you lean back to help them do more of the heavy mental lifting?

 Looking at my responses and current teacher disposition, my learning goals include…

Bridging Knowledge and Action

Understanding climate science, the changes, impacts, and opportunities associated are a few of the first components needed to support informed decision-making. We also know that learning about content and supporting students to move toward meaningful action are very different things however. Teaching using the socioscientific issues framework ensures that students tackle real-world issues and the ethical and social dimensions related, but we need to also tend to students' social and emotional well-being to help them process. Much of what students will learn regarding the climate emergency will bring out a wide variety of emotions that they will need help navigating. Exhibits 5.6 and 5.7 are interdisciplinary climate change lessons provided by incredible partners that foster students' emotional resilience in ways that center on their humanity and lived experiences.

Exhibit 5.6 The Power of Climate Storytelling in Empowering Classrooms

Learning with Jothsna Harris, Founder of Change Narrative

Climate storytelling is a transformative tool for educators seeking to cultivate empowered learning on climate change. Now a widely recognized strategy in climate communications for building public and political will for solutions, storytelling, when utilized in the classroom, supports academic learning by connecting to personal experience, which is necessary for meaningful dialog and action.

By creating spaces where students feel encouraged to share their unique perspectives, identities, cultural ways, lived experiences, and reimagined futures, educators can support the next generation of leaders to realize they are capable of rich climate dialog and decision-making. Because they have been able to articulate the unique lens from which they view it and, more importantly, to feel a sense of empowerment that comes from stepping into the power of their voice.

FIGURE 5.3 Change Narrative Educator Storytelling Workshop
Photo Credit: Michelle Garvey

The Intersection of Personal Narratives and Climate Education

The power of climate storytelling lies in the recognition that our understanding of climate change goes beyond the data and facts, because it is deeply personal and varied. Research tells us that information alone is not enough to change our behaviors. And, according to Neuroscientist, Antonio Damassio, "information is only meaningful to the extent that it evokes emotion."

Inviting students to share their reflections and experiences helps to reinforce scientific data, making it relevant and palpable. Whether recounting experiences of extreme weather events, grappling with eco-anxiety, or living in proximity to the cumulative impacts of environmental justice, each speaks to the depth of the human experience in a climate changed world, as well as the diversity of solutions.

When utilized in tandem, personal narrative, and climate education combines the head and the heart, and what emerges is empowered learning to realize self and collective efficacy, that what we do matters.

Cultivating Empowerment Through Expression

Empowering students to share their stories provides a platform to express and unpack intense feelings held about the climate crisis, and when shared together, it can give the sense that we are less alone. When students realize their voices matter, that they have a climate story and therefore something to contribute, they are more likely to engage in meaningful climate dialogs and decision-making. This empowerment extends beyond the classroom, encouraging students to participate in the broader climate movement. By acknowledging the power of their narratives, educators play a pivotal role in helping students envision themselves as integral contributors to the discourse surrounding climate change.

Inclusivity and Belonging: Honoring Identity and Culture

A crucial aspect of empowered learning in the classroom is its ability to create space for students to share their expertise by honoring diverse identities, cultures, and lived experiences, which illuminates the intersectionality of climate change. Climate change affects communities in distinct ways, influenced by geographical locations, systemic injustices, race, and socio-economic factors. It is critical to recognize and validate the reality of climate inequalities and that some narratives have been intentionally less visible than others in mainstream climate communications. Mainly Black, Indigenous, People of Color, people with disabilities, people who identify as LGBTQ+, and others who bear a disproportionate burden of climate impacts, yet whose perspective and proximity to the problem are imperative for just solutions. By acknowledging and amplifying frontline voices on climate change, educators can create an inclusive environment for learning and dialog across differences, essential for democracy.

Imagining a Bright Future: The Power of Shared Vision

Climate storytelling is not just about highlighting the challenges but also about exploring solutions. Encouraging students to share stories of resilience, adaptation, and community-driven initiatives transforms the narrative from despair to hope and action. This shift is crucial in motivating students to view themselves and their peers as capable of contributing to innovative solutions to the climate crisis.

Inviting students to imagine a vibrant and sustainable future is a key element of climate storytelling. By encouraging them to articulate their visions, educators empower students to speak into existence the world they want to inhabit. And, while climate stories are individual perspectives, together they form a collective narrative. This shared vision becomes a powerful catalyst, to bring desired futures into view and, importantly, making it more real for students themselves.

Crafting Climate Stories with Students

Ask students to reflect on their experiences of climate change and write aspects of their climate stories. You can use the following story prompts, or create your own. Stories can become tools for students to raise awareness and advocate for solutions, whether published in media, as a tool for climate conversations (with friends, family or even legislators), shared live at an event, or adapted into art, poetry or other forms of creative expression.

Storytelling Prompts

1. Think about a time in your community that made you smile, and reflect on that memory. Use visual imagery to describe the setting, the people, and aspects of your community that make it unique.

2. What is a time you felt your voice was powerful?
3. What is your experience of climate change? This could include direct impacts; wildfires, air quality, flooding, drought, heat, and extreme weather. Or how have you been thinking about climate change lately?
4. What is the vibrant and bright future you imagine is possible?

Students' writings may or may not connect, but they can form a strong story draft to work from. The following are tips as students refine their drafts and create a cohesive story.

Tips

♦ Can you identify the heart of the story? Typically this is the part of a story that includes a bit more vulnerability and emotion in sharing.
♦ Does the story draw the listener in through the following; sensory details (sight, sound, taste, touch, smell)
♦ Have you included climate facts to ground and reinforce your experiences?
♦ Is there an opportunity to introduce an ask, pose a challenge or an invitation into climate solutions?

Exhibit 5.7 Tending to Students' Social and Emotional Well-Being in Collaboration

Learning with Shraya Sharma, Content Manager at Empatico

Empatico and partners created the Empathetic Environmentalists program for educators around the world, which includes the lesson below. We hope that these experiences will help students to be successful in collaborating to overcome urgent world challenges that transcend national boundaries. To learn more, and gain access to more free educational resources, please visit https://empatico.buildersmovement.org/

FIGURE 5.4 Empatico Student Engaging in the Gratitude Nature Walk Activity
Photo Credit: Empatico

FIGURE 5.5 Students Collaborating During the Gratitude Nature Walk Activity
Photo Credit: Empatico

Activity Title: Gratitude Nature Walk

Overview: Living things grow and thrive when humans care for them. One way of introducing climate change to students is to inspire wonder during a gratitude nature walk. Encourage students to make observations and wonderings (questions) during the nature walk to build appreciation and gratitude. This sets the stage for students to think about their impact on the world around them, and instills a sense of responsibility for contributing to a more sustainable future.

Instructions:

1. Introduce the activity by asking students, "Does anyone know what climate change is?"

   Additional questions to draw out their funds of knowledge:

   ♦ What have you heard about it?
   ♦ Where have you heard about it?
   ♦ Who have you heard it from?
   ♦ How do you feel when you hear this phrase?

   ### Teacher Summary: What is Climate Change?
   Climate change (or global warming) is an urgent issue impacting our planet, and means that the temperature of our planet has been changing over a long period of time. Because of climate change, communities are experiencing (and will continue to experience) more extreme and unpredictable weather, leading to droughts, more acidic oceans, wildfires, and more high heat days. Climate change also affects the amount of rainfall, leading to different regions receiving more or less rain than they received in the past.

   We call it climate change rather than global warming because of the extreme changes in temperature (both hot and cold) and all these other things that happen as a result.

2. If students express feelings of anxiety, frustration, or sadness, consider guiding them through a mindfulness practice.

   ♦ You might say: "It's okay to feel all of these emotions when you think about climate change. Let's try a grounding exercise to help us focus on more positive feelings."

Lead students through the exercise by narrating the steps below:

   ♦ Sit or stand in a way that is comfortable for you, and imagine your favorite tree.

- ◆ Imagine your feet are roots growing into the Earth, anchoring you and connecting you to the world around you.
- ◆ Imagine your body is like the trunk of a tree, powerful and solid, providing protection and strength.
- ◆ Imagine your arms are like tree branches, strong but flexible, as you gently move them in the air.
- ◆ Imagine the warm Sun shining down on you, filling you with positive and calming feelings.

3. Next, explain to students that they will go on a gratitude nature walk around their school or home community. If students are engaging in a nature walk at home, always practice safety, and remind them to only go on nature walks with adult supervision.

- ◆ You might say: "We're going to go on a gratitude nature walk to notice things in the natural environment that we may feel grateful for. This can include the sight of falling leaves, seeing bright flowers sprouting, the sound of birds chirping, or anything else that you find beautiful or special! Then, we will make observations about human impact on our local community."

Educator Tip: If you are unable to take your students outside of the school, consider walking around the perimeter of the school and make observations around you, or use Google Earth to explore your community virtually if you have access to technology as another alternative.

4. Ask students to bring the "Gratitude Nature Walk" handout (www.bit.ly/gratitude_nature_walk) and something to write with. Guide students through their nature walk by sharing the following instructions:

- ◆ Notice any emotions that you experience, and what you can hear, see, or smell.

♦ Write, or draw pictures, on your handout to document any animals or plants that you notice.

♦ Look for any evidence of human impact on the environment, such as water usage from streams, land usage from urban development, or pollution.

♦ Write down any questions that you have about environmental impacts due to human activities.

5. After students complete their gratitude nature walk, engage them in a discussion by asking the following questions:

♦ Why are you grateful for the environment in our community? What do you love about it?

♦ What are some benefits that our environment provides to humans and other living creatures? (e.g., shelter, food, and water)

♦ How have humans impacted this environment?

♦ Should this space be protected for all the living creatures that depend on it (including us)?

♦ What is something that you wish people would do differently to help protect and respect our environment?

Encourage students to identify similarities and differences between their responses, emphasizing aspects such as gratitude, hopes for the future, and proactive ideas for a more sustainable future.

Purposeful Lesson Design Using ChatGPT

Are you ready to apply all that you've learned so far? I invite you to complete the activity in Exhibit 5.8 to explore the features of ChatGPT to teach about climate and the environment.

Exhibit 5.8 Accelerating Storyline Designs with ChatGPT

There was an article published by EdSource in 2021 to share about the challenges of PK-12 math education efforts in California. The writers posed a controversial theoretical question about whether PK-12 math should be removed because students can technically learn the skills and content with current technology and the internet. This article forced people to think about the purpose of math education, especially with such great advancements in technology. Let's pause for a second and considering the following questions:

♦ What if this article was written about *science* education?
♦ What if it was your class that was going to be canceled tomorrow?
♦ What would students really be missing that they couldn't get from the internet or ChatGPT?

As technology continues to advance, teachers will be needed more than ever to help students learn critical thinking skills, emotional resilience, and opportunities for locally relevant application. The more we come to rely on technology and efficient processing tools, the more we need to think about all the things that students cannot get from artificial intelligence that only we can provide. So let's *reframe*.

♦ When students look back at your class, what do they usually remember about your class? What do they share that they enjoyed most? What are their favorite moments?
♦ If your class was canceled tomorrow, what are major key topics, skills, concepts, or learning moments that students would miss out on that they could only get from *you* or *your class*?

The more we take advantage of ChatGPT or the internet as quick and efficient sources of information (careful because even these sources have bias and share misconceptions!), the more we can spend our time thinking about the importance of the previous questions (i.e. lightbulb moments, key skills, critical thinking, learning dispositions, increasing their emotional resilience, and more). ChatGPT can be an incredible planning tool to help you maximize on your learning and application, but remember that it's a language processing system and your knowledge, skills, and teacher disposition (i.e. qualities, beliefs, values, etc.) are needed to use it effectively.

Let's play with ChatGPT to apply what you've learned so far! Open up the free version of ChatGPT (sign in to save your entries for the future) and type in the following:

> You type: "I'm designing a multi-day lesson for my high school biology class to learn about keystone species, relevant ecosystems for a specific region, human impacts in both positive and negative ways, and it needs to be aligned to the NGSS science framework. Give me the scope and sequence for this type of lesson plan for Los Angeles, California specifically focused on the mountain lions and P-22."

ChatGPT initial generated response (pasted verbatim to support your learning): Scope and Sequence for High School Biology Lesson on Keystone Species: Mountain Lions and P-22 in Los Angeles, California

Day 1: Introduction to Keystone Species and Ecosystems

Objective: Introduce students to the concept of keystone species and ecosystems, with a focus on understanding the role of mountain lions as a keystone species in the Los Angeles area.

1. Engage: Begin with a brief discussion on what students know about ecosystems and species interactions. Introduce the concept of keystone species using examples from different ecosystems around the world.
2. Explore: Present information about mountain lions (Puma concolor) and their importance as a keystone species in ecosystems. Discuss the ecological role of mountain lions in maintaining biodiversity and ecosystem balance. Highlight the specific habitat and ecological niche of mountain lions in the Los Angeles area.
3. Explain: Provide background information on the ecosystem of the Santa Monica Mountains and its significance for wildlife conservation. Introduce P-22, the famous mountain lion living in Griffith Park, Los Angeles, as a case study for understanding human impacts on keystone species.
4. Elaborate: Engage students in a group activity or discussion to analyze how the loss of mountain lions could impact the ecosystem of the Santa Monica Mountains. Discuss potential consequences of disrupting the mountain lion population, including cascading effects on prey populations and plant communities.

Day 2: Human Impacts on Keystone Species

Objective: Explore the positive and negative human impacts on keystone species, with a focus on mountain lions and P-22 in the Los Angeles area.

1. Engage: Recap the previous day's discussion on mountain lions and their role as a keystone species. Introduce the concept of human impacts on ecosystems and wildlife.
2. Explore: Discuss the historical and current threats facing mountain lions in the Los Angeles area, including habitat loss, fragmentation, and human-wildlife conflict. Present case studies or examples of positive and

negative human interventions in wildlife conservation efforts.

3. Explain: Explore the specific challenges faced by P-22, including urbanization, habitat fragmentation, and the risks associated with crossing highways. Discuss ongoing conservation efforts aimed at protecting mountain lions and their habitats in urban environments.

4. Elaborate: Engage students in a debate or discussion on potential solutions for mitigating human impacts on mountain lions and other keystone species in the Los Angeles area. Encourage students to consider the role of policy, urban planning, and community involvement in wildlife conservation.

Possible Assessments: Assign students a project to research and propose solutions for addressing human impacts on mountain lions and other keystone species in urban ecosystems. Evaluate students' understanding through presentations, written reports, or multimedia projects that demonstrate their knowledge of keystone species, ecosystems, and human impacts in the context of Los Angeles, California.

Alignment with NGSS Science Framework:

Performance Expectations: HS-LS2–6, HS-LS2–7, HS-LS4–5
Disciplinary Core Ideas: LS2.A, LS2.C, LS4.D
Crosscutting Concepts: Cause and Effect, Systems and System Models, Stability and Change

You could stop there, but this book has given you more to work with to teach climate change in ways that activate agency and notice that the above is missing a great deal of opportunities. So let's *add more* to ChatGPT and see how we can bring this lesson up to our expectations to reflect our beliefs and values. Do *not* open a new ChatGPT window, just continue to type where you left off.

You type more: "Make sure to end the lesson with ways for my students who live in Los Angeles to take action to protect mountain lions, teach others about what has happened to the mountain lion population in creative ways that could include creating a poster or art exhibit, a social media campaign, or something creative to get people's attention."

After it gives you a "Day 3" addition to the lesson, continue typing based on what you've learned so far. Try the following:

You type more: "I want to teach this lesson in a way that leans into the socioscientific issues framework to develop students critical thinking skills, engages them in many opportunities for discourse, allows for them to argue from different perspectives, and leans into the ethical and social dimensions of the scientific issue. I also want to make sure that the lesson is grounded on equity and culturally relevant and responsive practices to meet the needs of all my students from diverse backgrounds. It's also important that students have opportunities to lead this lesson and build on their knowledge over time. I also would like to have an opportunity to explore and understand Indigenous history of the areas in which the mountain lions proliferated before, and how human activities and actions have decimated their current populations."

ChatGPT updated response (pasted verbatim to support your learning):

Day 1: Exploring Keystone Species and Ecosystems through a Socioscientific Lens

Objective: Introduce students to the concept of keystone species and ecosystems, while emphasizing critical thinking, discourse, and equity considerations.

1. Engage: Begin with a discussion on the interconnectedness of ecosystems and the impact of human activities on wildlife populations. Introduce the concept of socioscientific issues and their relevance to environmental conservation.

2. Explore: Present information on mountain lions as a keystone species and their role in maintaining ecosystem balance. Discuss the historical and cultural significance of mountain lions in Indigenous communities in the Los Angeles area. Explore how colonization and urbanization have impacted the habitats and populations of mountain lions and other wildlife.

3. Explain: Facilitate small group discussions where students analyze case studies or articles related to human-wildlife conflicts and conservation efforts in urban environments. Encourage students to consider multiple perspectives, including those of Indigenous communities, conservationists, policymakers, and local residents.

4. Elaborate: Provide students with opportunities to conduct research on the Indigenous history of the Los Angeles area and the relationship between Native peoples and wildlife. Encourage students to reflect on the ethical and social dimensions of human-wildlife interactions and conservation practices.

Day 2: Examining Human Impacts on Keystone Species and Equity Considerations

Objective: Explore the positive and negative human impacts on keystone species, while emphasizing equity, critical discourse, and ethical considerations.

1. Engage: Recap the previous day's discussions on keystone species, ecosystems, and socioscientific issues. Introduce the concept of environmental justice and its relevance to wildlife conservation.

2. Explore: Present information on the specific challenges facing mountain lions in the Los Angeles area, including habitat loss, fragmentation, and human-wildlife conflicts. Discuss the disproportionate impact of environmental degradation on marginalized communities and Indigenous peoples.

3. Explain: Facilitate debates or role-playing activities where students represent different stakeholders (e.g., developers, conservationists, residents) and argue their perspectives on wildlife conservation issues. Encourage students to consider equity considerations and the ethical implications of conservation decisions.

4. Elaborate: Invite guest speakers from local Indigenous communities or environmental justice organizations to share their perspectives on human impacts on wildlife and strategies for equitable conservation. Engage students in a collaborative project where they develop proposals for addressing human impacts on mountain lions while prioritizing equity and cultural relevance.

Possible Assessments: Evaluate students' understanding and critical thinking skills through class discussions, written reflections, and project presentations that demonstrate their ability to analyze socioscientific issues and consider diverse perspectives. Provide feedback on students' engagement in discourse, their ability to communicate effectively, and their consideration of equity and ethical dimensions in conservation efforts.

Alignment with NGSS Science Framework:

• Performance Expectations: HS-LS2–6, HS-LS2–7, HS-LS4–5
• Disciplinary Core Ideas: LS2.A, LS2.C, LS4.D
• Crosscutting Concepts: Cause and Effect, Systems and System Models, Stability and Change

By integrating socioscientific issues, equity considerations, and Indigenous perspectives into the lesson, students are encouraged to critically engage with complex environmental issues and develop a deeper understanding of the social, cultural, and ethical dimensions of wildlife conservation.

Remember the socioscientific issues framework stresses the importance of launching the lesson with the topic first, with anchoring or investigative phenomena to

support the design process as well. So try one more input into ChatGPT!

You type: "What are investigative phenomena that I could use to launch this lesson to start this off as an investigation into something engaging that is happening that isn't easily explained about the environment and the mountain lions? Are there short video clips, demonstrations, or activities that I could start with to help generate driving questions from students?"

ChatGPT updated response (pasted verbatim to support your learning):

To launch the lesson and engage students in an investigation, you can introduce anchoring phenomena that pique their curiosity about the environment and mountain lions. After introducing the investigative phenomena, facilitate a class discussion to generate driving questions and spark curiosity about the environmental issues surrounding mountain lions in the Los Angeles area. Here are some ideas for phenomena and corresponding activities:

1. Video Clips:

    ◆ Show a short documentary clip highlighting the story of P-22, the famous mountain lion living in Griffith Park, Los Angeles. Focus on his journey through urban landscapes and the challenges he faces.
    ◆ Share footage of wildlife cameras capturing mountain lions or other wildlife in unexpected urban areas, sparking discussion about human-wildlife interactions.

2. Interactive Maps:

    ◆ Use interactive maps to show the historical range of mountain lions in California and how it has been fragmented over time due to human development.
    ◆ Display maps illustrating recent sightings or conflicts between mountain lions and humans in the

        Los Angeles area, prompting students to investigate patterns and potential causes.

3. Role-Playing Activity:

- ◆ Divide students into groups and assign each group a role (e.g., mountain lion, developer, conservationist, resident) to simulate a town hall meeting discussing the presence of mountain lions in urban areas.
- ◆ Encourage students to research their assigned perspectives and prepare arguments for or against coexistence with mountain lions, fostering critical thinking and empathy.

4. Wildlife Tracking Demonstration:

- ◆ Demonstrate how wildlife tracking devices work and how scientists use them to study the movements and behaviors of mountain lions.
- ◆ Set up a mock tracking exercise where students use maps and data to track the movements of a virtual mountain lion, analyzing its habitat preferences and potential conflicts with human activities.

5. Indigenous Perspectives:

- ◆ Invite a guest speaker from a local Indigenous community to share traditional stories or knowledge about mountain lions and their significance in Native culture.
- ◆ Engage students in a discussion about the connections between Indigenous perspectives on wildlife and contemporary conservation efforts.

6. Mystery Scenario:

- ◆ Present students with a hypothetical scenario, such as a recent increase in mountain lion sightings in a suburban neighborhood, and challenge them to investigate possible explanations and solutions.

♦ Encourage students to ask questions, gather evidence, and propose hypotheses to solve the mystery, fostering inquiry-based learning and critical thinking skills.

I received all the inputs above (and more that I didn't paste here to keep it short) in less than 5 minutes. Amazing right?

Consider the following:

1. What other entries would you add to ChatGPT as you design that is aligned to your beliefs and values about teaching science?
2. How might you leverage AI and technology to help you design lessons and apply what you've learned?
3. Did you notice the limitations of ChatGPT and where your expertise, teacher disposition, and skills were needed to create a lesson that you would be proud to try out?

Reflections and Regrounding

Take some time to return to your teaching beliefs and values as you anchor your storylines around climate change to activate students. Think about why this work is important to you. In what ways does anchoring your science instruction on climate change reflect your beliefs about teaching and learning? How might it reflect the needs of teaching and learning for addressing the problems of today? What are teaching tools or approaches that you can readily take back to the class, and which ones are more challenging? Who might benefit from this type of instruction? Who can you collaborate with to teach this in interdisciplinary ways? Looking ahead, Chapter 6 will push you to think about and carefully integrate more ethical and social dimensions of climate change. This crisis impacts everyone and everything on Earth, and there's no denying that some are more impacted than others. As young people rise up around the world to lead the climate crisis movement, teaching them about the complex and political nature of climate change prepares them to be the changemakers we need.

Collective Voices for Climate Change Education

Jothsna Harris, Obama Leaders USA Fellow and Founder of Change Narrative LLC

In my role as a mother and my work in climate storytelling, I have witnessed the transformative influence educators hold. I admire your ability to shape empowered learning experiences and carve mean-ingful paths for students, nurturing identities and passions in a gen-eration whose necessary climate leadership holds the potential to build new worlds. In doing so, I invite you to lean into the inner work of discovering your own climate story, as an example. I am confident that you will deepen the personal lens for why climate change matters to you, feel the power of your voice when you share it, and to work from that place to spark agency in the classroom and beyond.

Emily Walker, Education Coordinator at EARTHDAY.ORG

Critical thinking, empathy, and resiliency are foundational skills students need to tackle climate topics, especially those around climate justice. Building these skills within our students is integral for students to become true stewards of the planet. We must remember that the cli-mate reality is their present and future. We can give students the tools and strategies that will ultimately open the door for change.

Additional Teacher Resources

Consider how the NGSS DCI builds across grades and subjects as you select phenomena—www.Bit.ly/DCIMATRIX
How the NGSS calls out SEPs by standard—Bit.ly/ARGUING
Learn more about the scientific practices of explaining and argumentation—Bit.ly/NGSSESA
Learn more about phenomena-based teaching—thewonderofscience.com
Learn more and see examples of storylines at Dr. Reiser's site—Bit.ly/BRSFTORYLINE

References

Kim, N. Y. (2021). *Refusing death: Immigrant women and the fight for environmental justice in LA*. Stanford University Press.

National Academies of Sciences, Engineering, and Medicine. (2018). *How people learn II: Learners, contexts, and cultures*. Washington, DC: The National Academies Press. https://doi.org/10.17226/24783

National Research Council. (2005). *How students learn: Science in the classroom*. Washington, DC: The National Academies Press. https://doi.org/10.17226/11102

Newmann, F. M., Smith, B., Allensworth, E., & Bryk, A. S. (2001). Instructional program coherence: What it is and why it should guide school improvement policy. *Educational Evaluation and Policy Analysis*, 23(4), 297–321.

6

Education as the Catalyst for Climate Action

Read this when:

- ♦ *You want to support students communicating about climate change outside the classroom.*
- ♦ *You're ready to engage students in lessons that integrate climate, environmental, and social justice.*
- ♦ *You're looking for ways to empower students to be agents of change in their communities.*

There is a real urgency and need to amplify our efforts as science educators to take on climate change for the sake of future generations. This work is challenging and time-consuming, but it can also be inspiring and serve as another way to excite students in STEM. Remember that students are bombarded with claims on the climate crisis every day, and rely heavily on teachers for scientific literacy skills needed to fight against disinformation campaigns and fake news. This final chapter will provide additional resources and tools to help you reflect on your teaching practices and ideas. Here you will find ways to help students get involved in taking action, which is just as important as teaching about the crisis itself. Rather than the "Gloom and Doom"

DOI: 10.4324/9781003478584-10

approach that leaves students feeling disempowered, you can be intentional with opportunities to build their capacity as agents of change in their own communities. The socioscientific issues framework calls for teachers to address the ethical and social dimensions of climate change in order to help students access their agency as scientifically literate citizens. By providing them with a safe space to explore climate change, you are helping them to see science as a tool they can use to arm themselves against misinformation.

Tips on Talking about Climate Change

It is no surprise that students will experience a great deal of skepticism and push back when communicating about the climate crisis. This segment provides accessible entry points on how to talk about this politically controversial topic that aims to debunk common misconceptions. In general, it is essential to consider the following points when talking about climate change: (1) People need to know how climate change impacts them directly in order to care, (2) they also need to know how their daily actions connect directly to the impacts of climate change in order to reflect on their decisions, (3) the conversation must be hopeful and solutions based, and (4) they need actionable steps to take if they are thinking of making a change based on what they have learned. Complete the activity in Exhibit 6.1 to help students have successful conversations about climate change.

Exhibit 6.1 Listening with Intent Activity

 Take it back to the class to help your students practice having important conversations about the climate crisis with friends and family members. Have students practice with one

another in class to help them become more confident. Provide the following guidance for their consideration:

1. **Have positive presumptions**—Assume that the person you're talking to means well and they are not trying to attack you personally with their opinions.
2. **Listening with intent**—Turns out that people are generally not good listeners. If you find yourself trying to jump in constantly to talk then you're not listening. You're not listening while you talk. Let the person know you hear them even if they're wrong, so that way they give you the same courtesy when you speak about something they might disagree with.
3. **You are not there to change or convince this person**—Your job is not to change this person's values or beliefs through this conversation. If they are skeptical about climate change, chances are they have been for many years. Remember that one conversation with you is not going to change that, but multiple non-threatening conversations might.
4. **Ask questions for understanding**—One thing you can do is help them find holes in their logic, data, or reasoning. Ask questions about where they found this information and ask if you could look it up together. Ask about how they learned about that information and that you're just trying to understand a different perspective. Reaffirm that you're not there to change the person.
5. **Plant seeds of doubt**—The thing about science is that it can be reproduced anywhere in the world by anyone taking the same steps. When you hear misconceptions or incorrect data usage, don't jump at the person with data and facts. Look up that information together and take the stance that you are just trying to learn as well.
6. **Engage in detective work**—What are their underlying values and beliefs about climate change? Is there a bigger reason for why this person may not want to

accept the unrefuted data about climate science? A good scenario to pose is the following:

a. *If you went to the doctor and they said that they ran your blood test at over 150,000 clinics and the results show that you have cancer. Would you believe the test results? What if they told you that they met with a panel of expert doctors treating cancer patients and they were 97% confident that you have lung cancer. Would you reject their diagnosis, refuse treatment, and pretend you don't have cancer?*

This scenario essentially describes the scientific consensus around climate change that students should have already learned about at this point.

7. **Resources at the top of your head**—Make a mental note of the following resources as you look up information with the person you're speaking with. It's okay if you don't have all the answers, but you should know where to find answers from credible organizations.

a. Our Climate Our Future—*ourclimateourfuture.org/*
b. 5 Gyres—*5gyres.org/*
c. Skeptical Science—*skepticalscience.com/*
d. Reach out to trusted adults and sources of knowledge in your life for support.

Consider the following teaching resources to help students communicate about climate change successfully.

◆ Alliance for Climate Education—Teaching youth to have conversations with parents about climate change summary report (www.Bit.ly/ACETALK).
◆ Our Climate Our Future—Show students this video to help them talk about climate change with others (www.Bit.ly/OURCLIMATE).
◆ Bending the Curve Book—Chapter 3 provides information on how climate change directly impacts human health and in disproportionate ways (www.Bit.ly/BTCBOOK).

◆ Watch Katharine Hayhoe's Ted Talk on communicating about climate change—(www.Bit.ly/CCTEDTALK)

Big Problems Require Big Solutions

We must recognize that the time to disrupt science education is now if we are to protect the planet for future generations. As Paulo Freire (2000) points out, the educational system can be the vehicle to transform society if we recognize what isn't working and all the ways it fails to support every student in accessing their agency. Although not every student will commit to a STEM job or career path, the goal is to ensure they are scientifically literate and informed decision-makers with options to pursue and persist in the field should they choose that path. This requires that we deliberately move away from the old standards and acknowledge the huge paradigm shift in how science should be taught and reflected in the curriculum. It is incredibly hard work to dismantle an institution, but institutions are created by people and their beliefs, and **both can change**.

It's currently the year 2024, and climate experts warn that we have less than 10 years to take action before we reach critical tipping points that will have devastating impacts for all life on Earth (Intergovernmental Panel on Climate Change [IPCC], 2018). The future they warn us about is **here** as explosive wildfires ravage across countries, prolific hurricanes wipe away entire cities, ocean acidification is threatening the food supply of almost 1 million people, the melting permafrost is causing sea-level rise and coastal erosion, and so much more. Young people across the country are already mobilizing for the climate crisis, and we need to get behind them to propel them forward as climate warriors. Leveraging the research-based pedagogical approaches and climate science standards in the NGSS, take time to revisit your teacher dispositions and the potential of what science education can be to meet this moment.

Make no mistake that the news and social media headlines will continue to become more dire and worrisome as we near 2030 (the year we need to cut our emissions worldwide by half) and then 2050 (the year we need to achieve net-zero carbon emissions globally). It's essential to remember that **we determine the next points on the carbon emissions graph with our individual and collective actions, and every degree is worth fighting for.** My inspiration comes from colleagues like Dr. Brian Tarroja, a senior scientist from the UC Irvine's Advanced Power and Energy Program. Brian currently works on evaluating and creating sustainable solutions for large clean energy systems and approaches. He reminded me that each passing year has the potential to bring about the change needed to bend the curve. When 2030 and 2050 comes and goes (which it will do naturally), it's important that educators understand and continue to communicate that **it's not over.** Every degree that we keep down is worth every effort we can give, and in Brian's words,[1]

> There's no threshold where everything's okay and you go above it and everything just goes really really bad. **Every reduction is worth fighting for.** There's no threshold where people should just give up. Yes, there's 2°C, but that doesn't mean that if we go to 2.1°C that everyone should give up. 2.1°C is better than 2.2°C, and 2.2°C is better than 2.3°C. It's a continuous effort to see what we're able to do.

Learning From Other Educational Leaders

In June 2020, CA State Superintendent Tony Thurmond, along with Dr. Pedro Noguera (Dean of The School of Education at USC), Sujie Shin (Deputy Executive Director of CCEECA), and Dr. Daryl Camp (President of CAAASA) gathered to talk about preparing students for the future amidst the pandemic. Their panel session focused mainly on shifting the paradigm for how educators teach underrepresented minority (URM) students in low-income communities and the growing inequities due to distance learning. When we think of culturally relevant and responsive teaching, Dr. Noguera noted that teachers need to recognize that it will take more than the first week of school to learn about students' deep cultures, backgrounds, interests, and ways to effectively engage them. More importantly, they pointed out that schools with resources still fail at educating URMs in staggering numbers (see the Beyond the Schoolhouse Report at www.Bit. ly/uclaBTSH).

We cannot ignore the fact that historically marginalized communities made of predominantly URMs suffer more from the consequences of climate change, than affluent high-income communities that have the means to survive in increasingly challenging conditions. Although this panel session was not to address climate change directly, there were many points made that ran parallel such as (1) a paradigm shift is needed for educators to teach differently than what they experienced themselves, (2) we need to teach in culturally relevant and responsive ways to engage students in issues that directly impact them, and (3) equity in the classroom is an intentional act by teachers that can disrupt traditional ways of teaching and learning. As an intersectional problem, climate change requires more than one solution and diverse ways of thinking. This complex issue requires all stakeholders to understand that an intersectional approach will bring about more meaningful, transformative, and systemic changes.

Climate & Social Justice

We need courageous leadership for climate mobilization. As teacher leaders, we can elevate the voices of young people who are rising up to create solutions for climate and environmental injustices in their communities. It is essential to create change that people can **feel** especially when the injustices involve access to clean air, water, food, land—or education unveiling that the lack of the former should not be accepted or normalized. For climate change, teaching through a social justice lens means that teachers embed opportunities for students to learn more about (and make decisions on) the ethical and political dimensions of the climate crisis. Remember that students will be bombarded with messages about climate change every day whether or not teachers decide to provide time to unpack these claims (Caranto & Pitpitunge, 2015; Carter & Wiles, 2014; Hansen, 2010; Hestness, McDonald,

Breslyn, McGinnis, & Mouza, 2014; Hodson, 2003; Matkins & Bell, 2007; Somerville & Hassol, 2011). If we choose to omit this part from our science curriculum (although the NGSS SEPs call for meaningful discourse opportunities to engage in argumentation from evidence, making claims, etc.), we are choosing the safety of silence on issues that impact students. We also miss opportunities to prepare them to engage as young leaders who are both scientifically literate and informed-decision-makers for society.

Move from Equity-Centered to Justice-Oriented

The first thing teachers need to understand is that there are no strategies or a checklist for equity-center or justice-oriented teaching. It's about reflecting on your teaching practices, values, and beliefs and making intentional decisions to support every learner in your class. Consider asking yourself the following questions to begin unpacking your practices:

1. Think of a famous scientist. Who is this person? What is their cultural background? Do other famous scientists look like this person as well?

   Reimagined—Do the scientists that students learn about look like or identify with them? Is there value in having students learn about underrepresented minorities in STEM fields as they develop their own STEM identities? Should we make an intentional effort to expose students to diverse scientists? Is there value to that?

2. Think of another famous scientist. How did you learn about this person? What does this person sound like? Do all scientists use the same language? Do all scientists express ideas in the same way?

   Reimagined—Who decides what scientists should sound like when they speak? When students express ideas in your class, do you notice *what* they are saying (content) or *how* they are saying it first (behavior)? Have you ever corrected the way a student expresses their ideas? What are you using as a guide when you have students rephrase or restate their response? Are there cultural ties to your view of who a scientist is, what they sound like, or their weight of their contributions?

3. Think of your science classroom. What takes up the most space on the walls of your room? Would an observer be able to identify your teaching values and beliefs in that space?

   Reimagined—In what ways might students see themselves represented in the classroom? Is that important as they develop their STEM identity? Are you consistent in making efforts to unveil and learn more about your students' deep culture (what they believe is right versus wrong, etc.)? How might you elevate students' voices, ideas, and contributions so they feel valued in that space?

4. Think about the cultural backgrounds, multifaceted identities, and lived experiences of your current students. How do your daily interactions with them affirm who they are? What might that look like when designing learning experiences that allow for students to bring their wholeselves to the classroom? What are the cues you pick up on that a student is bridging knowledge from home and the community with expanded knowledge from the lesson?

   Reimagined—What are the cues you pick up on when students begin to lean into wanting to explore the ethical or social dimensions related to the topic you're teaching? How does your body and mind react to those instances (when you're fully aware)? How are students' desires to connect what they learn in class with what they hear beyond class connected? How do you determine when to pause, and/or change course in your lesson to accommodate their wonderings (especially if they're starting to question the ethics or implications of injustice related to the topic)?

As you contemplate how science teaching could look, sound, and feel like with the questions above, Figure 6.1 will help you to better understand the social justice spectrum to see where you are and where you want to be.

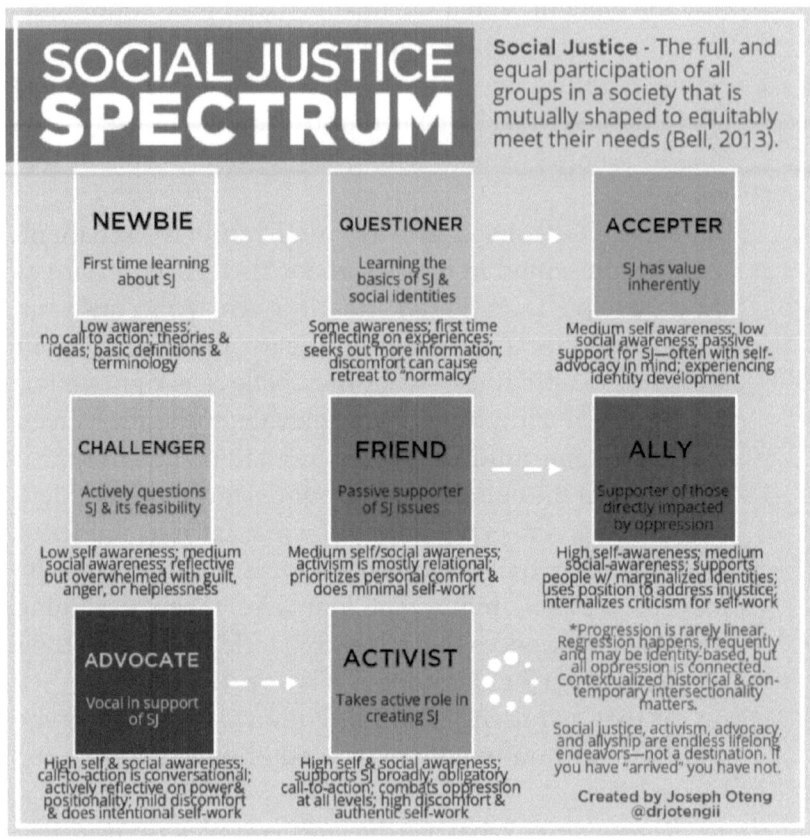

FIGURE 6.1 The Social Justice Spectrum
Image Credit: Dr. Joseph Oteng @drjotengii—Social Justice Educator

Incorporating Social Justice Issues into Your Storylines

As previously mentioned, those who contribute the least to global climate change, will be impacted the most. It is essential to dismantle oppressive and discriminatory systems that allow for things to continue to operate in the same way to fulfill the status quo. A powerful way to get students to think about climate change issues is to make it directly relevant to them. When teaching about energy, teachers can tie in interactive tools (such as EPA TRI) to show where power plants are in the community as well as what chemicals are often released into the community soil, air, and water (and noting where it is not commonly found). Another example is when teaching students about aqueous solutions or water, teachers can start with statistics (i.e. Water Crisis at *Water.org*) to show how nearly one billion people lack access to clean water each year or that every two seconds a child will die from a water-related disease (and how this basic human right will become a bigger challenge due to climate change). These examples highlight a systems-thinking approach to learning about science to reveal how interconnected issues are and their direct human impacts.

As the keynote speaker at CSTA (2018), Dr. Ram Ramanathan strongly urged teachers to prepare students as climate warriors for this uphill battle to improve the planet they unjustly inherited. He argues that he calls them climate warriors because, "this is a fight for survival to protect the habitat for generations yet to be born (2018)." Students need to be taught about the science of climate change so that they can develop solutions we desperately need to bend the curve. Students need to learn about the inequities of the world and in their communities, and they need to remain hopeful. It is essential that educators provide tools, resources, and support to channel students' desire to help better the world using science. Table 6.1 offers a few ideas that can be incorporated as anchoring or investigative phenomena that allow for students to learn science through a climate and social justice lens (Looking ahead, Table 6.3 lists organizations that build student agency around similar issues).

United Nations Sustainable Development Goals (SDGs)

The United Nations SDGs were written to unite the world in addressing major goals that range from improving health, education, and inequity, to taking on the climate crisis and mitigating its impacts. Teachers can use the SDGs to center curricula around real-world issues and connect with educators around the globe working toward the same goals. Figure 6.2 identifies all 17 sustainable development goals that teachers can explore more in depth and find teaching resources for. Exhibit 6.2 is an example lesson that leans in SDGs 13 and 14.

TABLE 6.1 Climate and Social Justice Issues with Phenomena

Climate & Social Justice Issues	Possible Phenomenon
Air Pollution and Impact on Specific Communities	Use this resource to analyze air pollution by looking at the "pollution burden" score and health scores by zip-code. Generate student questions, and find ways to move to civic community engagement based on the revealed data. *Resource: www.Bit.ly/CAtool*
Clean Water and Impact on Specific Communities	Use these resources to reveal contaminants in water by zip-code. Are there ways to protect communities from these contaminants? In what ways might education on water quality lead to community action based on the data? *Resource: www.Bit.ly/CAtool or https://www.fractracker.org/map/us/*
Environmental Justice and City Responses	Explore your city's Climate Action Plan (CAP) to see how they define and address environmental justice implications of climate change. *Resource: www.ca-ilg.org/climate-action-plans* *Specific Example:* Los Angeles City pLAn (see Chapter 1 Environmental Justice) https://plan.lamayor.org/
Fracking and Impact on Specific Communities	Use this resource that maps out each state and the influence that fracking has on water, land, and air quality in each community. What data can be drawn and analyzed from this site? How might education on fracking protect communities in the future? *Resource: https://www.fractracker.org/map/us/*
Heat Island Effect or Sea-Level Rise Impact on Communities	Use this resource to analyze high heat days by zip-code to show the heat island effect or sea-level rise in different communities. What will happen when there are power outages as the heat index increases and demands more energy? How are climate models showing sea-level rise and collapse of ecosystems in your area? How might education on sea-level rise protect communities in the future and lead to better city planning? Use this to spark student conversations and move to community action. *Resource: https://cal-adapt.org/*
Hospital Fatality Rates and Impact on Specific Communities	Using information provided on hospital malpractices to determine hospital grades, what patterns can be seen in lower rated hospitals compare to highly rated ones? What can we do with this information and move towards community action? *Resource:* www.Bit.ly/HOSPITALGRADES
Housing Inequities and Impact on Specific Communities	Using Images of low-ses and high-ses communities to show disparities, take questions, explore implications, and provide discourse opportunities. *Resources:* www.Bit.ly/PHOTOINEQUALITY *Specific Neighborhood Example Resource:* UCLA's "Beyond the Schoolhouse" report for Los Angeles www.Bit.ly/BeyondTheSchool.
Wildfire and Air Quality Data Tracker	Use this resource to analyze recent changes in air quality by zip-code. Have students analyze the graphs and anomalies they identify. Have them zoom out to look for wildfires nearby and data trends. *Resource: https://fire.airnow.gov/*

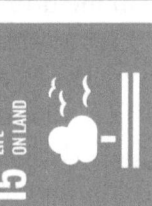

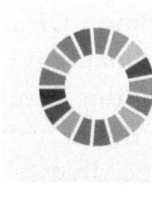
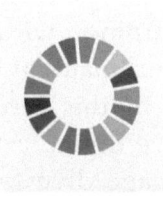
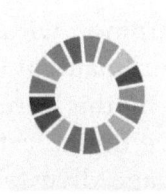

FIGURE 6.2 United Nations Sustainable Development Goals

Image Credit: https://www.un.org/sustainabledevelopment/ (The content of this publication has not been approved by the United Nations and does not reflect the views of the United Nations or its officials or Member States).

Exhibit 6.2 Lessons That Integrate Art to Inspire Action

Learning with Dr. Jennifer Cao, Special Programs Coordinator at ECCLPs

In this introductory lesson, educators will use creative digital artwork from the *Ocean Decade Exhibition* to engage diverse learners in ocean literacy, aligning with Sustainable Development Goals 13 and 14. The Ocean Literacy Framework is a set of principles designed to support the understanding of the ocean's role in shaping the Earth's systems and its influence on human life. Students will understand the fundamental importance of the ocean and its interconnectedness with humans through creative imagery.

Access the Supporting Teaching Slides: www.bit.ly/ TEACHOLP

Introduction: Students will apply their prior knowledge and funds of knowledge to articulate the importance of the ocean including the benefits of the ocean to humans.

Exploration of the 7 Things Everyone Should Know About The Ocean: Students will participate in a round-robin activity where they will discuss their initial reactions to digital art. Students will rotate every three minutes between digital artworks and use the provided discussion prompts to guide their conversations.

Discussion Group Prompts

◆ How did the artwork make you feel?
◆ What's your first reaction to the artwork?
◆ Does this artwork make you wonder about anything?
◆ How does the artwork represent how humans and the ocean are connected?

Then educators will display the following Ocean Literacy Principles (OLPs) to engage students in whole group discourse on which digital artwork represents which OLP.

 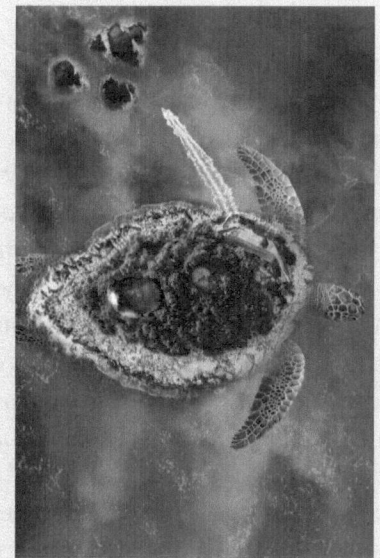

Credit: The Ocean Agency - Artwork by Umut Reçber

Credit: The Ocean Agency - Artwork by Aanish Peshave

- The Earth has one big ocean with many features.
- The ocean and life in the ocean shape the features of Earth.
- The ocean is a major influence on weather and climate.
- The ocean makes Earth habitable.
- The ocean supports a great diversity of life and ecosystems.
- The ocean and humans are inextricably interconnected.
- The ocean is largely unexplored.

Students are strongly encouraged to provide rationale for their claims. To support, educators can provide additional information about each OLP while revealing the accompanying digital artwork.

These creative digital pieces of art lower the barriers to learning for students to dive into life below water. The following lessons in the series will spotlight topics including mangroves, coral reefs, and seagrasses, highlighting the ocean's critical role in mitigating climate change and supporting biodiversity. Ocean imagery *toolkits* enhance student understanding of these concepts.

Resources That Build Student Capacity

When Ricky Gervais was being interviewed by late night show host Stephen Colbert in 2017,[2] he defended his stance about the need to trust science and the process of science in very relatable terms. He shared that if we were to burn any work of fiction today, 1,000 years from now that work will cease to exist. If we were to burn all the peer-reviewed high-quality scientific research or studies explaining the world around us, however, they would still be found true even 1,000 years from now with the same results regardless of who it's authored by or where on Earth. We need to arm our students with resources, tools, and approaches they can use to respond confidently when confronted by a climate change skeptic. This complex problem requires people to work together to actively find solutions. This segment will provide teachers with pedagogical approaches to support students through co-constructing knowledge together.

Dictogloss Activity

Dictogloss is a classroom activity where students learn to use one another as resources of information to reconstruct a short text. This activity positions all students as capable learners and contributors of information through collaborative efforts. Students also learn quickly that they need each other in order to complete the task because everyone can contribute. The following is an overview of how to facilitate the activity with your students.

1. Start with a current event, scientific article, or science concept with multiple parts. Let the students first analyze on their own to find big ideas, identify questions they have, and try to make sense of it. Then provide time so they can share through discourse their level of understanding with an elbow partner.
2. Next, go over the ideas as a class and let the students know they get one chance to write down or draw as many details as possible.
3. Have students share what they were able to write or draw with each other, and their level of understanding.

4. Let them know that you'll be going over it one last time and their job is to piece together the current event, article, or science concept with as many details as possible. This time, tell them they can work in groups of three or four to come up with a plan for the second listen.

5. After the last exposure, allow for them to work together to reconstruct the reading with as much details as possible. Then give them time to share any issues they might have encountered, preventing them from fully understanding the reading/concept.

6. Facilitate a discussion about how this activity emphasizes the need to work together and to use one another as resources of information. Emphasize how this positions them as scientists because these researchers rarely work in isolation and complete studies in teams to help see the full scope of the problem.

More Science Teacher Talk Moves

Consider using talk moves for productive conversations in your class (Refer back to Chapter 1– Exhibit 1.8). Whether you are facilitating whole group or small group discourse, think about how equitable those experiences are for students. Do you find that students tend to engage in dialog with you as their teacher, instead of their peers? When they share their thoughts and ideas during whole group discourse, are they confident in participating and feeling open to having their ideas challenged? Consistently using talk moves supports students to talk with each other to help them engage in the process of science as a community of learners. Scientists communicate with one another often and they also have to think about how they will share their research in accessible ways. Build up your students' abilities with the following resources:

1. STEM Teaching Tools Talk Flowchart—stemteachingtools .org/brief/35
2. Talk Moves Goals Reference Sheet—Bit.ly/TALKMOVES
3. Exploratorium Science Talk Tool—Bit.ly/ExploratoriumST

Introducing Students to CER

Since argumentation is a skill students need to become more informed-decision-makers in society, teachers can introduce CER using media and technology to help students learn the general process. See Figure 6.3 for an example provided by a student showing their change in thinking over time, which resulted in changes to their driving question and initial claim.

Claim, Evidence, Reasoning (CER) Worksheet

My Question:

My question: How do the white Rhino become extinct?

Driving Question: *Write additional questions you have as you learn more in a different color.

What factors caused the white Rhino to go specifically extinct?
What factors cause species, like a white Rhino, to go extinct?

Claim: Why do you think this is happening?	White Rhinos are extinct because of human impact such as the people's needs for ivory and hunting.
Evidence: What science content, information, or data supports your claim? What would you use to convince someone that your claim should be adopted? This can be in bullet points.	(handwritten notes, partially legible) Biosphere includes all living things on Earth and is connected to all spheres. Invasive grass hazard near highway — cars will light in middle of highway — grass will light. Droughts = more fire b/c dried vegetation. humans are at fault for intentional fires. people demand = companies increase supply by destroying habitats → destroying habitats = more pandemics due to mosquitoes increase. 6th mass extinction: what do we mean? who will it impact? humans are also species. extinctions. Some pipes. overfishing. 5 extinctions (diff. causes but same ↑ CO₂). CO₂ — atmosphere, ocean - ocean acidification. Top signs pointing to mass extinction: 1) biological annihilation, 2) species are threatened w/ extinction → 1 million species, 3) insects are declining → bees & pollination = food source, 4) amphibians decline, 5) extinction domino effect → ecosystem depend on each other, 6) natural habitats are shrinking = species ↓, 7) some animals have become too little = no role in ecosystem
Reasoning: How does the class information, data, or content support or make you want to adopt, modify, or reject your current claim (Write modified claims under your initial claim)? How does the data or information explain why the phenomenon is happening?	I would modify my claim to where it is more generalized to more species. Humans are heavily in the role of animal extinction. In various natural habitats, humans are creating intentional fires that destroys ecosystems. However, this only happens because regular consumers are increasing their demands to everyday packaged products. For example, chips and other packaged foods required palm oil in which companies set various areas on fire to create more land for supply. In different parts of the sea that used to have many sea life are now overrun by sewerpipes and overflowing to sustain ourselves. The extinction domino effect when one part of the ecosystem becomes extinct and disrupts the ecosystem causing other animals/organisms to go extinct. In addition, some animals, like koalas, are functionally extinct in which there are no longer apart in an ecosystem and will soon go extinct.

(See next page for more evidence) |

FIGURE 6.3 Iterative Student CER Worksheet Sample

CER SESSION 7
More evidence:
· 2 Billion people are estimated to become climate migrants by 2100.
· Alicia Key-let me in
  - war → affecting unarmed citizens
  - citizens fleed to Mexico → crossing the border
  - families separated
  - more than half refugees are children
· heat index
  - heat + humidity
  - heat kills ppl
  - ↑ humidity + heat = ppl overheating (Deadly)
  - 125°F Chicago = 700 ppl died
  - 165°F + humidity = internal organs fail
· Carbols
  my address: 55-60%
    Pollution Burden 87   could be from
      PM2.5 66   freeways and constructions?
    Population 5,282
· Air conditioning brings more heat and makes the area hotter than it should be

I would modify my claim to be even more specific. Human actions can negatively impact ecosystems of plants and animals by causing them to go extinct But these actions also impacts ourselves. In various natural habitats, humans are creating intentional fires that destroys ecosystems. However, this only happens because regular consumers are increasing their demands to everyday packaged products. In different parts of the sea that used to have many sea life are now overrun by sewerpipes and overflowing to sustain ourselves. Human actions are heavily involved with the climate crisis. As from what we have learned from the heat index, our atmosphere is rising in temperature and mixing the heat w/ humidity, our internal organs will fail and kill us. It has been estimated that 2 billion people will become climate migrants in 2100. However, there are some parts of the world in which isn't open to the idea of opening their borders. In addition, if countries do open borders, it will also effect the resources, animals, and people currently living in that place.

FIGURE 6.3 (*continued*)

To give you a high level view of the learning process, I am going to provide details of how teachers from the climate change education program learned to use CER more effectively. It is important to position yourself at the level of your students to effectively anticipate student responses, what they know, what they notice, and what they might be curious about. Exhibit 6.3 is a breakdown of how other teachers learned about class facilitation of CER.

Exhibit 6.3   Learning from Other Teacher's Experiences

Background: The following program was provided in 2018 and 2020 across four non-profit environmental organizations in Los Angeles to secondary science teachers. Teachers from over 40 school districts were represented and participated in 21 hours of professional development on climate change education for teachers.

Program Day Three—Introducing teachers to CER

1. Using a CER worksheet, teachers were prompted to view a clip and write down the claim of the clip.
2. Teachers then watched a Super Bowl Kia commercial showing actress Jenny McCartney driving around the world saving the planet in her new hybrid car.
3. Teachers then shared with each other what they believe the claim of the clip is. They were brought together as a group to share ideas out loud. Just as a person watching the advertisement at home, different people will generate different claims for what they are presented with.
4. Teachers were prompted to re-watch the clip again to collect evidence to support their individual claims. They were advised to write down anything they felt served as evidence to support their initial claim.
5. When the clip ended, they took turns sharing the evidence gathered to support their original claim.
6. To help unveil teachers' rationale, they were asked to listen to each other and then adopt one claim that had the most credible evidence presented. The teacher will notice at this point that students begin to evaluate their own credibility of evidence.
7. As they engage in argumentation using evidence, they are given sentence frames to help guide them if needed.
8. Lastly, they engaged in the reasoning portion by evaluating the strength of evidence presented (if any). They had to decide to accept/modify/or reject their initial claim as a result.

Teacher Tip: Super Bowl commercials make for good introductions to argumentation in science using evidence. Popular social media events also work such as (1) The NASA Broomstick Challenge, (2) BBC Documenting Rare Flying Penguins, or (3) Liquid Mountaineering.

Teachers expressed concern about facilitating the "reasoning" part with students because they had difficulty doing so in the past. With CER, students will tend to reiterate what they already wrote as "evidence" to also serve as their "reasoning." I modeled for teachers how to unveil what students' current understandings might be in order to connect new information with what they believe is true. It is essential to unveil students' current understandings to address any misconceptions they may have before introducing new content. Exhibit 6.4 are questions that teachers can pose during whole-class discourse to uncover students' reasoning on evidence selected.

Exhibit 6.4 Evaluating Credibility of Evidence Using CER

You can pose the following questions to help students think more critically about the evidence they selected to support their claim.

1. What do you know about the topic currently?
2. Why did you select that as evidence?
3. Is your evidence subjective?
4. Is it possible that someone in this class disagrees with your evidence?
5. Can you find credible and vetted data to support your evidence?
6. What is the source of your data?
7. Is what you selected actually considered evidence?
8. Would you classify your evidence as weak or strong? Why?
9. How does the evidence support your initial claim? If it doesn't, how can you modify your claim to reflect the evidence you have?

Imagine if your students firmly believed in a misconception for several years. As teachers, we must understand that it will take more than a couple of class sessions to debunk their misconceptions. Worse yet, if the misconception is never unveiled then teachers operate under the assumption that students will just adopt new ideas and abandon their long held beliefs. They may memorize the content for the test, but walk out with the same misconceptions they came in with. Transforming teaching means taking the time to know what students know so that they can have more depth of knowledge. Students must have opportunities to question what they know and how they came to know it to bridge the gap between old and new ideas.

To help students resist scientific misinformation online, Tumblehome created teacher-friendly lessons that are easily adaptable. Once students get comfortable with the consistent CER process in class, this resource provides teaching resources that push for students to use technological research skills to identify fake news. It is easily adaptable for the needs and level of your students and can serve as an introductory lesson on the importance of argumentation in science using evidence. Exhibit 6.5 is a list of topics that students can research using tools provided by Tumblehome as a good introduction to politically controversial climate change content.

Exhibit 6.5 Identifying Fake News Research Activity

Have students work in small groups to find out if they are victims of fake news. Using the "Resisting Scientific Information" teaching resources and tools, have them research their topic online. Before they look up the answers to their topic, have them engage in CER to generate a group question and a claim for their question. Once they are ready, allow them to research more about their topic online while writing down questions they may have. Lastly, have groups report out their findings and whether they accept/reject/modify their initial claims in light of new evidence.

Possible Topics (Can be Current Events or Graphs to Interpret as Well)

1. Analyze the Heartland Institute Book for Teachers online denying climate change.
2. News highlighting that everyone is now breathing in microplastics.
3. Jellyfish causing the emergency closure of nuclear power plants.

Exploring "I Don't Know"

On the Cult of Pedagogy, Connie Hamilton shared a useful tool during her interview to help teachers to explore what students mean when they say, "I don't know (IDK)." Turns out that when students respond with, "IDK" there could be a variety of reasons that move beyond not knowing the answer to your question. Are they saying, "IDK" because they don't understand the question? How helpful would it be for a teacher to hear the student say instead, "I didn't understand the question, can you rephrase it for me?" How might the teacher respond knowing the exact point of confusion rather than dealing with "IDK" broadly? Hamilton also notes how powerful it can be to just add the word "yet" to their response to let them know that they are still growing and that it's okay to not know yet. She states that when students don't know something we need to transition them to knowing more about that thing, **rather than terminate that line of thought**. I recommend downloading the resource on the website and providing a copy to each student to utilize when they are engaging in discourse. Simply asking if they can have more time to think about the question to process the ideas, reveals so much more to you as a teacher than IDK.

 *Take it back to the class with this resource to help facilitate this learning segment www.Bit.ly/HandlingIDK.*

Gauging Student Progress

Mini-Lab Portfolios

Creating mini-lab portfolios allow you to enhance the traditional lab reports and position students as scientists. Students can complete this using an online format (Using Google Slides, Keynote, Prezi, etc.) or a physical mini-presentation board using recycled manilla envelopes. Remember to scaffold instruction on writing these segments so students are not overwhelmed and are working with support. It is generally a good idea to have the first portfolio completed in lab groups of 3–4, then the second portfolio in pairs, and eventually completed individually.

See Appendix F for example student portfolios and the template to use with your students. Access these online at www.EmpoweredScienceTeachers.com (Book Resources → Google Drive)

Students as Experts

Once your students have gotten used to creating mini-lab portfolios, you can push them further by sharing what they have learned with the community. Have students present their portfolio to one adult on campus in under five minutes. Allowing students to teach their history, math, or art teacher about science builds their confidence because they are the experts in that moment. Consider opening up the criteria to any adult on campus so they can also speak to a wide variety of individuals (i.e. security guards, nurses, counselors, administrators, custodial workers, etc.). Have the adult listener submit a quick Google Survey evaluating the students' presentation so you can provide them with feedback afterwards. Normally one week is enough time to complete this activity and encourage students to look at the Google Survey prior to presenting so they can anticipate the questions. It is helpful to send out an email to your administrator to let them know that students will be reaching out to adults across campus so that students receive the support needed.

Access these resources online at www. EmpoweredScienceTeachers.com (Book Resources → Google Drive)

Science Oral Defense

Gauging students' level of understanding can also be completed through an oral defense at the end of your instructional segment. This is an alternative method of assessing students by allowing them to present their knowledge through discourse with their peers. Students are provided guidelines prior to the defense so they can prepare their 10-minute segment. In groups of 3–4, one student will present their big ideas with supporting evidence while the other students listen, take notes, and evaluate the strength of information/evidence provided. The grade is determined by the average of all the listeners and your final evaluation as you circulate and listen. Allowing students to explain what they feel are the big ideas and what serves as evidence is another way to support different learners and varying needs while engaging in the Process and Nature of Science.

Access these resources online at www. EmpoweredScienceTeachers.com (Book Resources →️ Google Drive)

STEM Teaching Tools

If you are looking for ways to better align your assessments to gauge student learning on a deeper level, consider exploring these resources from the *STEM Teaching Tools*. This database has online professional development and a variety of resources to integrate three-dimensional learning. For grades 6–12, the first resource allows you to integrate the Cross Cutting Concepts (Bit.ly/CCCQUESTIONS), and the second resource allows you to integrate Science and Engineering Practices (Bit.ly/SEPQUESTIONS). These resources help to modify current questions for assessments, while providing pocket questions to pose for class discussion around any topic.

Take it back to the class with this resource to help facilitate this learning segment at http://stemteachingtools.org/.

Interactive Tools to Enhance Instruction

Climate change content needs to be directly relevant to students' lives to show the importance of why humans must take action now to mitigate the impacts and bend the curve. These interactive tools allow teachers to use data to highlight the impacts on students' communities (many by zip code). Position students as drivers of instruction by having them use, interpret, discuss, and pose questions they have about what they are seeing. Students can analyze by city, state, or country, which allows them to unveil the larger societal and systemic inequities or injustices through the data. Recall that 100kin10 predicted that environmental and climate justice issues will move to the forefront of education because of student demand. These tools provide the historical context needed to understand those issues they care most about. Table 6.2 outlines top interactive tools that teachers can use to engage students in science practices to see direct community impact.

Activating Students for Action

Remember that climate science should not be taught in isolation, represented over a few disconnected lessons, or taught without opportunities to explore possible solutions. We cannot engage students in science practices when lessons are not grounded in real and culturally relevant phenomena. The following segment provides teachers with resources on supporting students to take action in their own communities as capable scientists or engineers. Consider taking risks to provide intentional opportunities for students to explore the ethical dimensions of climate change to also support them as young community leaders. Lastly, move away from instruction that disempowers students.

Community Science Experiences

You can have students support research efforts in a variety of ways by integrating community science experiences in your curriculum. These programs allow students to collect or analyze data for researchers in their community. The data is often used to inform decision-makers and students can be part of that experience to learn more about the process of scientific research. These programs are also great ways to position students as capable scientists where they can be experts in the process and share what they learn with those around them. Explore Table 6.3 to connect with organizations across the nation that build student agency.

TABLE 6.2 Incredible Interactive Educational Tools

Online Tool & Access	Description & Uses
CalEnviro 3.0 OEHHA—California Office of Environmental Health Hazard Assessment *Access:* www.Bit.ly/CAtool	Based on the census tract, students can look up their neighborhood's population burden score. The higher the overall number, the higher the burden for that population. It also measures toxicity in the air, water, and soil as well as social and demographic information by zip code.
Climate Interactive—Free tools to simulate/model solutions for climate change Access: www.Bit.ly/CCINTERACTIVE	Have students engage with models to analyze and interpret data for NGSS and to engage in science practices. These models will inform students' understanding through their programs: En-ROADS, C-ROADS, and ALPS.
EPA TRI—US Environmental Protection Agency Toxic Release Inventory *Access:* www.Bit.ly/EPATOOL	This online interactive tool allows students to look up what toxic chemicals are released in their neighborhood by zip code. They can see what chemicals are released into the air, water, and soil as well as facilities in your neighborhood.
Fire and Smoke Map—Interactive map by zipcode *Access:* www.fire.airnow.gov/	This online interactive tool lets students analyze fires by state and zipcode to analyze air pollution quality within the last week. Type in the zipcode, hover over the censor (dot), ask questions, zoom out for larger analysis and to generate more questions.
NOAA Data in the Classroom—Interactive Tools *Access:* www.dataintheclassroom.noaa.gov/	Explore a variety of interactive tools to enhance student learning in the areas of coral bleaching, sea level, ocean acidification, El Nino, and Water Quality. Each tool comes with teacher lesson plans and student worksheets.
Population Earth—Teacher Resources *Access:* www.Bit.ly/POPEDUCATION	Explore this resource to teach about the current population numbers and use data to introduce concepts like carrying capacity, shortage of resources, human impact on Earth, etc.
Prezi Climate Change—Teaching Lessons *Access:* www.Bit.ly/PreziCC	Explore a variety of lesson slides on climate change content, and their video platform that allows you to create virtual lessons as well.
The New York Times—What's Going on in This Graph? *Access:* www.Bit.ly/NYTGRAPHS	Experts in a variety of fields (including climate change) post graphs once a week to engage students across America in analyzing and interpreting data.

TABLE 6.3 Organizations That Build Student Agency

Online Tool & Access	Description & Uses
5 Gyres—*Students can be community ambassadors.* Access: https://www.5gyres.org/ ambassadors	Students will have access to resources to host community events as an ambassador and ways to connect with others across state lines. There are also research studies that take place in different states every year to help lawmakers make informed decisions using data collected by community/citizen scientists.
350—Connect with like-minded people internationally and get involved in your own community. Access: https://350.org/ get-involved/	Join the international movement of people trying to end fossil fuels for a sustainable world. *"We believe in a safe climate and a better future—a just, prosperous, and equitable world built with the power of ordinary people.—350"*
Climate Generation—Connect with students across the nation taking action on climate change. Access: https://www.climategen. org/	Teachers can access free curriculum resources and information on professional development opportunities. Students are also able to act on climate change by joining the Youth Environmental Activists network.
Ocean Conservancy—Be an ocean advocate and take action. Access: https:// oceanconservancy.org/ action-center/	Students can become involved as advocates of the ocean by taking action on marine life, protecting the Arctic, climate change, and the health of the ocean.
Our Climate Our Future— Promote student activism and networking. Access: https:// ourclimateourfuture.org/	Get students involved in community activism and help them to connect with like-minded students across the nation on the topic of climate change. Teachers also have free access to teaching resources.
Strategic Energy Innovations (SEI)—Teacher curriculum and student certification programs for future green careers. Access: https://www.seiinc.org/	Explore free NGSS-aligned science teacher curriculum on topics such as energy, climate change, and biomimicry. Students are also able to earn certification for building solar cells, completing an energy audit, and others to build up their resumes for future green jobs.
Sunrise Movement—Connect with like-minded people internationally and get involved in your own community. Access: https://www. sunrisemovement.org	Join the movement of young people creating a greener future and bending the curve on climate change. *"Together, we will change this country and this world, sure as the sun rises each morning.— Sunrise Movement"*

Online Tool & Access	Description & Uses
The Biomimicry Institute—Have students engage in the Youth Design Challenge. *Access:* https://biomimicry.org/	TBI prides itself on supporting students to develop skills, be curious, and gain new perspectives through their educational resources. *"Biomimicry is a practice that learns from and mimics the strategies found in nature to solve human design challenges—and find hope along the way.— The Biomimicry Institute"*
The Years Project—Everyone can take action in various ways with TYP. *Access:* https://theyearsproject.com/	Explore accessible ways to take action in your community and connect with fellow teachers or students. They also have teaching resources (video stories, clips, lessons, etc.) for teachers through their climate classroom portal.
Understanding Global Change—an interactive tool to support systems thinking among students https://theyearsproject.com/	Embed this tool into your lesson plan to help students see how larger systems are affected and at different scales.

 *Take it back to the class and check out Dr. Lê's second Padlet dedicated to community science programs at Bit.ly/ CommunitySci*

Teaching to Catalyze Climate Action

Thank you for your time, dedication, and commitment as an educator and an influential agent of change. We have the huge responsibility and privilege to teach about the science of climate change in ways that are sobering, hopeful, and empowering. As young leaders across the world mobilize to demand action on climate change, we can build their skills as informed decision-makers and solution designers with science. We need to prioritize consistent and culturally relevant opportunities that allow students to take ownership of their own learning experiences every day. Have confidence that this will push them to access their creativity, passion, innovation, and resilience to help bend the curve.

We Are Here Now in the Middle of the Climate Crisis, but Make No Mistake That We Also Get to Determine Where We Go Next as a Collective

As you move forward to anchoring your instruction around the ongoing climate crisis, remember that you are not alone in bringing climate change through a social justice lens to the forefront of education. Connect or create a coalition of educators to ensure you can give and/or receive support when needed (because it will be essential to avoid burnout and to build your emotional resilience through community). Reach out to your teaching colleagues within and beyond your school to see who is ready to take on the challenge with you. Explore local and national organizations that are doing something about the climate crisis through education and advocacy. Intentionally draw upon teaching practices and resources that build capacity in your students as young leaders and community change agents. Finally, recognize the rippling impact that you make as an educational leader. We are raising the next generation of climate and environmental stewards and leaders. **We will succeed because we can do this together and there's no better time to start than right now.**

Track your professional growth by re-examining your beliefs about teaching and learning.

Tracking Your Professional Progress	
Student Learning Experiences in Your Class	*On a scale of 1 (extremely disagree) to 5 (extremely agree), to what extent do I agree or disagree with the following statements? Why?*
Today's Date:	
Students often discuss policies related to science.	
Students often collect and analyze current data or information.	

Student Learning Experiences in Your Class	On a scale of 1 (extremely disagree) to 5 (extremely agree), to what extent do I agree or disagree with the following statements? Why?
Students often discuss ethical issues related to science.	
Students often co-construct knowledge with me during lessons.	
Students often drive the instruction in my class as capable contributors and doers of science.	
Students are often positioned as current scientists or engineers.	
Students often learn about Indigenous traditional ecological knowledge.	

Curriculum Design Elements	On a scale of 1 (extremely disagree) to 5 (extremely agree), to what extent do I agree or disagree with the following statements? Why?
Today's Date:	
I often build lessons/units around anchoring or investigative phenomena.	
I often present climate or environmental issues at the start of each unit or lesson.	
Students often engage in argumentation and making claims based on evidence.	
Students often engage in meaningful discourse opportunities.	
My lessons/units are often centered around real-world issues that are directly related to students' lives or community.	

Curriculum Design Elements	On a scale of 1 (extremely disagree) to 5 (extremely agree), to what extent do I agree or disagree with the following statements? Why?
Students often engage with the Nature of Science principles.	
Students often connect with local or relevant field experts and researchers.	

My Teacher Attributes	On a scale of 1 (extremely disagree) to 5 (extremely agree), to what extent do I agree or disagree with the following statements? Why?
Today's Date:	
I have to know everything about a particular topic before teaching it.	
I am extremely confident in my knowledge of climate science.	
I feel comfortable admitting to students when I don't know the answer to their question.	
I am comfortable teaching about open-ended issues where I cannot predict student responses.	
I am not the only source of knowledge on climate and the environment for students.	
I often experience imposter syndrome when teaching, even for topics that I have strong expertise in.	
I often reflect on my pedagogical decisions to improve my practice.	

 *Look over your responses and compare them to the results from the introductory chapter. What shifts in thinking have occurred for you throughout this book? What are some goals that you have now? What were the biggest takeaways from this resource? What support do you still need? What do you feel empowered to do next?*

Collective Voices for Climate Change Education

José Gonzalez, Founder of Latino Outdoors

As educators, practitioners, and movement builders, we know all too well that "Doom and Gloom" narratives can be limiting. While they can certainly prompt some action, they are still grounded in the space that leads and contributes to the climate anxiety that so many youth suffer from at the moment. I have said that to be "revolutionary" is to honor the root of the word, which is "cycles," and leads me to ask, "How can I be revolutionary by healing and repairing regenerative cycles rather than being complicit in destroying them?" Because it is when we break the regenerative cycles of the planet that we get in environmental trouble. This framing still grounds me in action, but it does it with a growth-oriented life logic—the same which invites the shift from "natural resources" to "natural relatives," and from succeeding by "functioning like a well-oiled machine" to "functioning like a well-nurtured meadow." This is the "Do and Bloom" rather than "Doom and Gloom" that I believe is possible, and which YOU play a critical role in. So let's go be revolutionary and thrive together like well-nurtured meadows.

Joana Tavares, Director of WLAC's California Center for Climate Change Education

Dear fellow educators, in our mission to raise climate change awareness and promote justice, we face daunting challenges. Yet, united in purpose, we gain strength from our community.

We must empower students to envision a sustainable future, understanding that climate issues intertwine with social, economic, racial, and gender injustices. By teaching these connections, we equip them to be holistic change-makers, addressing both environmental and societal disparities. In our collective determination, we inspire a generation to tackle climate change while striving for a fairer, more equitable world. In solidarity, Jo.

Additional Teacher Resources

Connect with others through the Alliance for Climate Education—acespace.org/

Explore this website that tracks water inequities and how to get involved—water.org

NGSS Tasks Pre-Screener & Screener—https://www.nextgenscience.org/taskscreener

Trust for Public Land Park Scores interactive tool—https://www.tpl.org/parkscore

STEM Teaching Tools Assessment Resources—https://stemteachingtools.org/tgs/Assessment

STEM Teaching Tools Equity Resources—https://stemteachingtools.org/tgs/Equity

Teaching About the United Nations Sustainable Development Goals—https://mcic.ca/uploads/public/files-sf/SF-Full-FINAL-WEB-ISBN-2021-EN.pdf

Notes

1 Dr. Brian Tarroja's remarks were taken from his 2023 interview recorded by the ECCLPs UCI Task Force assistants.
2 Actor Ricky Gervais' remarks were made in 2017 during The Late Show with host Stephen Colbert.

References

Caranto, B. F., & Pitpitunge, A. D. (2015). Students' knowledge on climate change: Implications on interdisciplinary learning. In *Biology education and research in a changing planet* (pp. 21–30). Singapore: Springer.

Carter, B. E., & Wiles, J. R. (2014). Scientific consensus and social controversy: Exploring relationships between students' conceptions of the nature of science, biological evolution, and global climate change. *Evolution: Education and Outreach*, 7(1), 6.

Freire, P. (2000). *Pedagogy of the oppressed*. 30th anniversary edition. New York: Continuum.

Hansen, P. J. K. (2010). Knowledge about the greenhouse effect and the effects of the Ozone Layer among Norwegian pupils finishing compulsory education in 1989, 1993, and 2005—What now? *International Journal of Science Education*, 32(3), 397–419.

Hestness, E., McDonald, R. C., Breslyn, W., McGinnis, J. R., & Mouza, C. (2014). Science teacher professional development in climate change education informed by the next generation science standards. *Journal of Geoscience Education*, 62(3), 319–329.

Hodson, D. (2003). Time for action: Science education for an alternative future. *International Journal of Science Education*, 25(6), 645–670.

IPCC. (2018). *Global Warming of 1.5°C. An IPCC Special Report on the impacts of global warming of 1.5°C above pre-industrial levels and related global greenhouse gas emission pathways, in the context of strengthening the global response to the threat of climate change, sustainable development, and efforts to eradicate poverty* [Masson-Delmotte, V., P. Zhai, H.-O. Pörtner, D. Roberts, J. Skea, P.R. Shukla, A. Pirani, W. Moufouma-Okia, C. Péan, R. Pidcock, S. Connors, J.B.R. Matthews, Y. Chen, X. Zhou, M.I. Gomis, E. Lonnoy, T. Maycock, M. Tignor, and T. Waterfield (eds.)]. In Press.

Matkins, J. J., & Bell, R. L. (2007). Awakening the scientist inside: Global climate change and the nature of science in an elementary science methods course. *Journal of Science Teacher Education*, 18(2), 137–163.

Somerville, R. C., & Hassol, S. J. (2011). The science of climate change. *Physics Today*, 64(10), 48.

Appendices

Appendix A – Interactive Science Mapping Tool

Using UC Berkeley's Online Mapping Tool

Access the tool at www.Bit.ly/CALTOOL2

1. First, give students an opportunity to review the different parts of the picture to select where they started in their investigation (i.e. they made an observation, asked a question, they had a curiosity, etc.).
2. Next, students can click on the bubbles on the image to fill out the left tabs that appear with relevant details to document their process and how it will change over time based on what they learn, observe, notice, continue to wonder.
3. The end result will look like the image below. Students can save their current maps to iterate on over time by clicking "Save" on the upper righthand corner to their computer or Google account.
4. Another option is to have students submit their Process of Science mapping tool as a PowerPoint to show all the steps they took and to unveil their thinking over time by selecting the button in the upper right hand corner with three gray bars. Select the PowerPoint option to download. The PowerPoint file will upload to the desktop. Students can upload to the learning management system for grading, use it as a presentation, or share it for peer feedback.

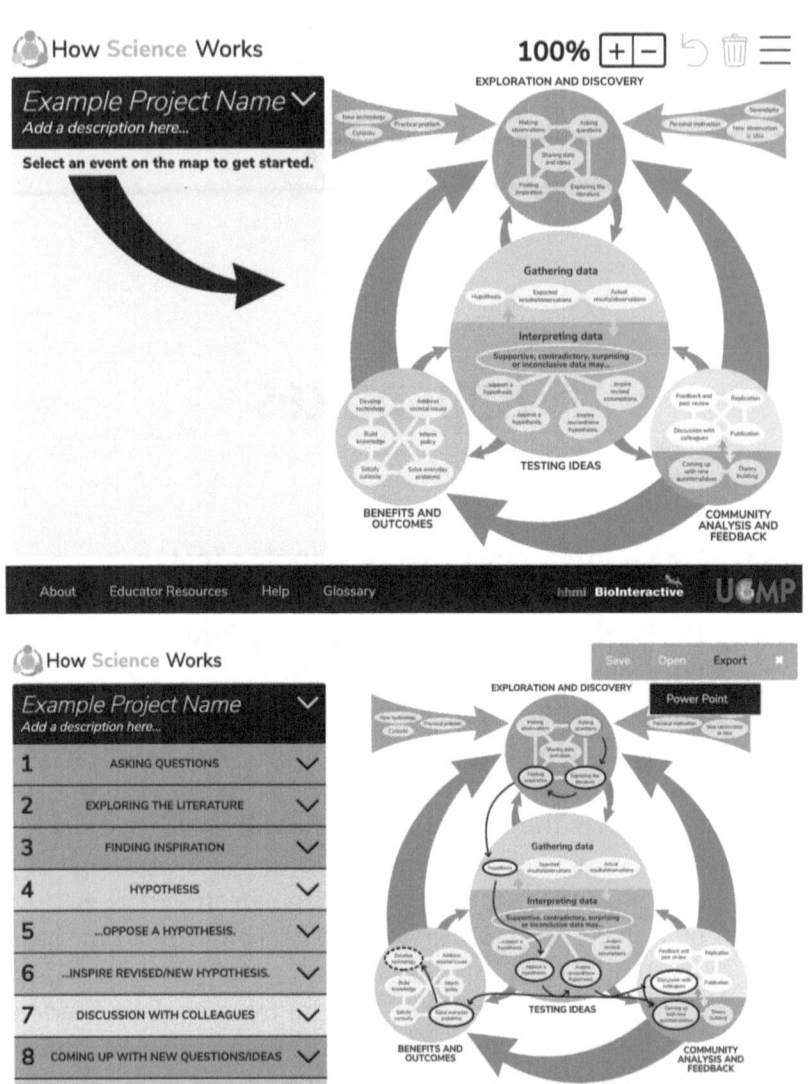

Appendix B – Curriculum Map Templates

To access printable versions of these templates to organize the information, please go to www.empoweredscienceteachers.com (under "Resources" tab).

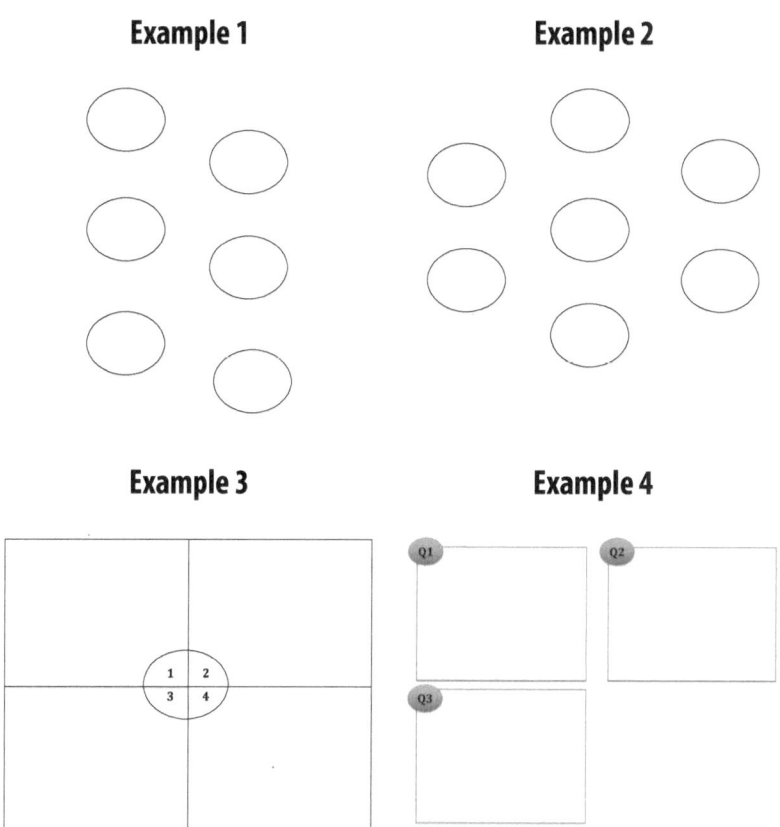

Example 1

Example 2

Example 3

Example 4

Appendix C – Sample Curriculum Maps

The following are curriculum map samples from participants in the climate education program showing their iterative design processes over the days (in the order of Biology, Chemistry, and Environmental Science).

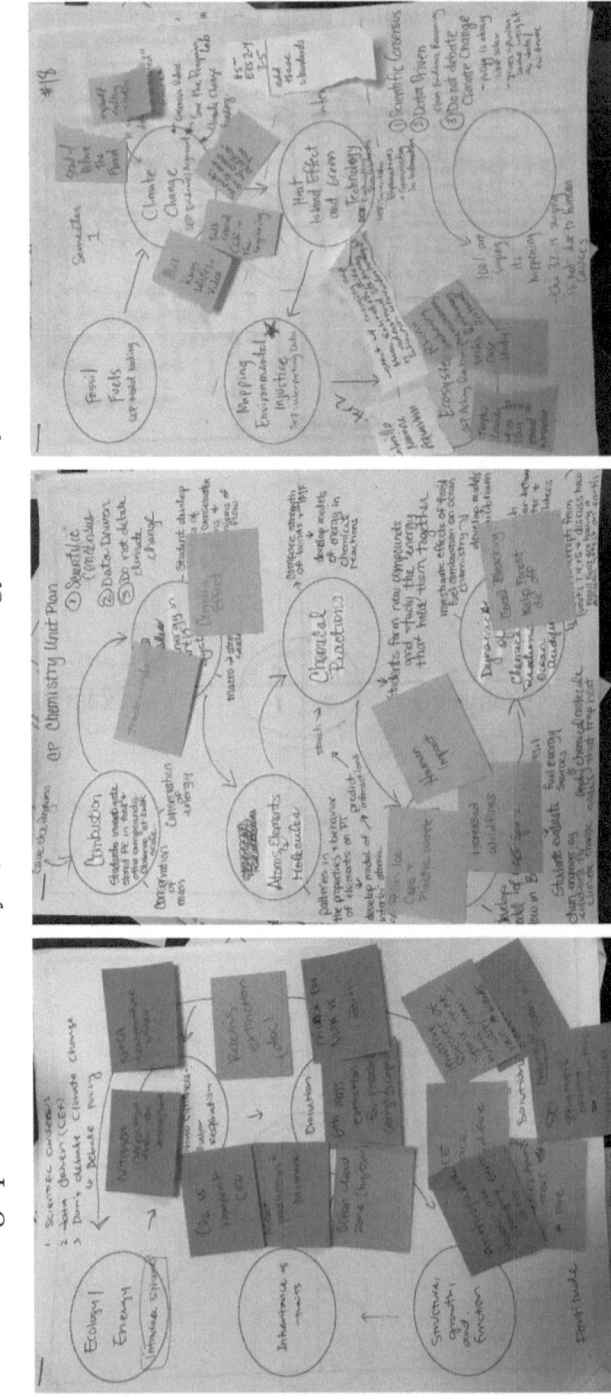

Appendix D – Developing the Driving Question

Developing the Driving Question ¿ ?

My Original Question

Phenomenon – Something that happens in nature that is not easily explained.

(Write a question about the phenomenon you just saw. What are you curious about?)

Group 1's Question:

Group 2's Question:

Group 3's Question:

Group 4's Question:

Group 5's Question:

Group 6's Question:

Group 7's Question:

Group 8's Question:

Final Anchoring Question:

Appendix E – CER Example Worksheet

Claim, Evidence, Reasoning (CER) Worksheet

My Question:

Driving Question: *Write additional questions you have as you learn more in a different color.*

Claim: *Explain what is happening and why you think that.*	
Evidence: *What science content, information, or data supports your claim? What evidence would you use to convince someone else that your claim should be adopted?*	
Reasoning: *As you continue to learn, modify and strengthen your current claim (Write modified claims under your initial claim).* *How might the information, data, or content you learn in class support your claim? How might the data/info. Help to explain why the phenomenon is happening?*	

Appendix F – CER Resource Worksheet

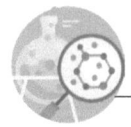 **Mini-Lab Portfolio Components**

1. Research Topic & Background
What were you looking at specifically for this lab? What are you trying to discover or learn about? Write 2 paragraphs about your lab topic, all the major ideas, any related content that you're applying, state standards that apply, and relate it to this specific lab. What are you research questions? What type of study is this?

2. Materials and Methods
List all materials used and your detailed procedures. This section needs to be explained in your own words to an audience of 7 year-olds (be very detailed about what you did and provide rationale for why you did those things).

3. Creative Title
If someone passed by, this title should captivate them to want to learn more about your topic (Keep in mind your audience and appropriateness).

4. Hypothesis
The hypothesis is written in the format of an "If," "Then," "Because" statement.

5. Abstract
Give a summary of your research topic and lab findings to help others determine if they want to read about your work (Think about what you are trying to find, how you found it, and what you found).

6. Photos and Results
This area should include photos from the experiments and any apparatuses that were used to conduct the experiments (can be drawn nicely if necessary). Include any data tables and graphs analyzing your results. You may attach photos of the lab set up, the process, and the results (before and after). This is where your statistical data should be displayed with mathematical work and results.

7. Discussion and Conclusion
Minimum of two paragraphs discussing your hypothesis (do you accept or reject it based on your results), sources of error that might affect your lab results, and science implications (what can you do with this information? What can be invented? How is this useful? Why should we care?). What major ideas/information will you have a better understanding of now that you've completed the lab? Explain and elaborate on why you think differently now and how. Any works cited can be listed anywhere on the poster.